Big Data, IoT, and Machine Learning

Internet of Everything (IoE): Security and Privacy Paradigm

Series Editor

Mangey Ram

Professor, Graphic Era University, Uttarakhand, India

IOT

Security and Privacy Paradigm

Edited by Souvik Pal, Vicente Garcia Diaz, and Dac-Nhuong Le

Smart Innovation of Web of Things

Edited by Vijender Kumar Solanki, Raghvendra Kumar, and Le Hoang Son

Big Data, IoT, and Machine Learning

Tools and Applications

Edited by Rashmi Agrawal, Marcin Paprzycki, and Neha Gupta

Internet of Everything and Big Data

Major Challenges in Smart Cities

Edited by Salah-ddine Krit, Mohamed Elhoseny, Valentina Emilia Balas, Rachid Benlamri, and Marius M. Balas

For more information about this series, please visit: https://www.crcpress.com/Internet-of-Everything-IoE-Security-and-Privacy-Paradigm/book-series/CRCIOESPP

Big Data, IoT, and Machine Learning

Tools and Applications

Edited by
Rashmi Agrawal, Marcin Paprzycki,
Neha Gupta

CRC Press
Taylor & Francis Group
Boca Raton London New York

CRC Press is an imprint of the
Taylor & Francis Group, an **informa** business

First edition published 2021
by CRC Press
6000 Broken Sound Parkway NW, Suite 300, Boca Raton, FL 33487-2742

and by CRC Press
2 Park Square, Milton Park, Abingdon, Oxon, OX14 4RN

Library of Congress Cataloging-in-Publication Data

Names: Kolawole, Michael O., author.
Title: Electronics : from classical to quantum / Michael Olorunfunmi Kolawole.
Description: First edition. | Boca Raton, FL : CRC Press, 2020. | Includes bibliographical references and index. | Summary: "This book discusses formulation and classification of integrated circuits, develops hierarchical structure of functional logic blocks to build more complex digital logic circuits, outlines the structure of transistors, their processing techniques, their arrangement forming logic gates and digital circuits, optimal pass transistor stages of buffered chain, and performance of designed circuits under noisy conditions. It also outlines the principles of quantum electronics leading to the development of lasers, masers, reversible quantum gates and circuits and applications of quantum cells"-- Provided by publisher.
Identifiers: LCCN 2020011028 (print) | LCCN 2020011029 (ebook) | ISBN 9780367512224 (hardback) | ISBN 9781003052913 (ebook)
Subjects: LCSH: Electronics. | Electronic circuits. | Quantum electronics.
Classification: LCC TK7815 .K64 2020 (print) | LCC TK7815 (ebook) | DDC 621.3815--dc23
LC record available at https://lccn.loc.gov/2020011028
LC ebook record available at https://lccn.loc.gov/2020011029

ISBN: 9780367336745 (hbk)
ISBN: 9780429322990 (ebk)

Typeset in Palatino
by Deanta Global Publishing Services Chennai India

Contents

Section I   Applications of Machine Learning

Section II   Big Data, Cloud and Internet of Things

Preface

INTRODUCTION

Big data, machine learning and the Internet of Things (IoT) are the most talked-about technology topics of the last few years. These technologies are set to transform all areas of business, as well as everyday life. At a high level, machine learning takes large amounts of data and generates useful insights that help the organisation. Such insights can be related to improving processes, cutting costs, creating a better experience for the customer or opening up new business models. A large number of classic data models, which are often static and of limited scalability, cannot be applied to fast-changing, fast-growing in volume, unstructured data. For instance, when it comes to the IoT, it is often necessary to identify correlations between dozens of sensor inputs and external factors that are rapidly producing millions of highly heterogeneous data points.

OBJECTIVE OF THE BOOK

The idea behind this book is to simplify the journey of aspiring readers and researchers to understand big data, the IoT and machine learning. It also includes various real-time/offline applications and case studies in the fields of engineering, computer science, information security, cloud computing, with modern tools and technologies used to solve practical problems.

Thanks to this book, readers will be enabled to work on problems involving big data, the IoT and machine learning techniques and gain from experience. In this context, this book provides a high level of understanding of various techniques and algorithms used in big data, the IoT and machine learning.

ORGANISATION OF THE BOOK

This book consists of two sections containing 13 chapters. Section I, entitled "Applications of Machine Learning" contains six chapters, which describe concepts and various applications of Machine Learning. Section II is dedicated to "Big Data, Cloud and Internet of Things", and contains 7 chapters, which describe applications using integration of Big Data, cloud computing and the IoT. A brief summary of each chapter follows next.

Section I: Applications of Machine Learning

Chapter 1 on "Machine Learning Classifiers" deals with the fundamentals of machine learning. Authors facilitate a detailed literature review and summary of key algorithms concerning machine learning in the area of data classification.

Chapter 2 on "Dimension Reduction Techniques" discusses the dimension reduction problem. Classical dimensional reduction techniques like principal component analysis, latent discriminant analysis, and projection pursuit, are available for pre-processing and dimension reduction before applying machine learning algorithms. The dimension reduction techniques have shown viable performance gain in many application areas such as biomedical, business and life science. This chapter outlines the dimension reduction problem, and presents different dimension reduction techniques to improve the accuracy and efficiency of machine learning models.

Application distribution platforms, such as Apple Store and Google Play, enable users to search and install software applications. According to statistica.com, the number of mobile application downloaded from the App Store and Google Play has increased from 17 billion in 2013 to 80 billion in 2016. Chapter 3, "Reviews Analysis of Apple Store Applications Using Supervised Machine Learning", contains a case study, which aims at building a system that enables the classification of Apple Store applications, based on the user's reviews.

Biomedical research areas, such as clinical informatics, image analysis, clinical informatics, precision medicine, computational neuroscience and system biology, have achieved tremendous growth and improvement using machine learning algorithms. This has created remarkable outcomes, such as drug discovery, accurate analysis of disease, medical diagnosis, personalised medication and massive developments in pharmaceuticals. Analysis of data in medical science is one of the important areas that can be effectively done by machine learning. Here, for instance, continuous data can be effectively used in an intensive care unit, if the data can be efficiently interpreted. In Chapter 4, "Machine Learning for Biomedical and Health Informatics", a detailed description of machine learning, along with its various applications in biomedical and health informatics areas, has been presented.

Chapter 5, "Meta-Heuristic Algorithms: A Concentration on the Applications in Text Mining", presents a detailed literature review of meta-heuristic algorithms. In this chapter, 11 meta-heuristic algorithms have been introduced, and some of their applications in text mining and other areas have been pointed out. Despite the fact that some of them have been widely used in other areas, the research, which shows their application in text mining, is limited. The aim of the chapter is to both introduce meta-heuristic algorithms and motivate researchers to deploy them in text mining research.

Deep learning is used within special forms of artificial neural networks. When using deep learning, user gets the benefits of both machine learning and artificial intelligence. The overall structure of deep learning models is based upon the structure of the brain. As the brain senses input by audio, text, image or video, this input is processed and an output generated. This output may trigger some action. In Chapter 6, "Optimising Text Data in Deep Learning: An Experimental Approach", challenges related to deep learning have been discussed. Moreover, results of experiments conducted with text-based deep learning models, using Python, TensorFlow and Tkinter, have been presented.

Section II: Big Data, Cloud and Internet of Things

Data and analytics technology trends will have significant disruptive effect over the next 3 to 5 years. Data and analytics leaders must examine their business impacts and adjust their operating, business and strategy models accordingly. Chapter 7, "Latest Data and Analytics Technology Trends That Will Change Business Perspective", details these trends and their impact in businesses.

In Chapter 8, "A Proposal Based on Discrete Events for Improvement of the Transmission Channels in Cloud Environments and Big Data", a method of data transmission, based on discrete event concepts, using the MATLAB software, is demonstrated. In this method, memory consumption is evaluated, with the differential present in the use of discrete events applied in the physical layer of a transmission medium.

With the increasing number of sensing devices, the complexity of data fusion is also increasing. Various issues, like complex distributed processing, unreliable data communication, uncertainty of data analysis and data transmission at different rates have been identified. Taking into consideration these issues, in Chapter 9, "Heterogeneous Data Fusion for Healthcare Monitoring: A Survey", the authors review the data fusion algorithms and present some of the most important challenges materialising when handling Big Data.

Chapter 10, "Discriminative and Generative Model Learning for Video Object Tracking", is devoted to video object tracking. For video object tracking, a generative appearance model is constructed, using tracked targets in successive frames. The purpose of the discriminative model is to construct a classifier to separate the object from the background. A support vector machine (SVM) classifier performs excellently when the training samples are low and all samples are provided once. In video object tracking, all examples are not available simultaneously and therefore online learning is the only way forward.

Chapter 11, "Feature, Technology, Application, and Challenges of Internet of Things", includes varied features, technology, applications and challenges of the Internet of Things. An attempt has been made to discuss key features of IoT ecosystems, their characteristics, significance with respect to different challenges in upcoming technologies and future application areas.

The sustainable smart city is an attempt to fulfil the philosophy, which employs an integrated approach toward achieving environmental, social and economic goals, through the use of ICT. Chapter 12, "Analytical Approach to Sustainable Smart City Using the IoT and Machine Learning", explores different enabling technologies and, on the basis of completed analysis, proposes an analytical framework for a sustainable smart city using the IoT and machine learning.

Finally, Chapter 13, "Traffic Flow Prediction with Convolutional Neural Network Accelerated by Spark Distributed Cluster", addresses challenges in the area of flow prediction, by deploying an ApacheSpark cluster, using CNN (Convolutional Neural Networks) model for training, when the data consists of images captured by webcams in New York City. To efficiently combine CNN and Apache Spark, the chapter describes how the prediction model is re-designed and optimised, while the distributed cluster is fine tuned.

We sincerely hope that readers will find this book as useful as we envisioned it when it was conceived and created.

Rashmi Agrawal, Marcin Paprzycki, Neha Gupta

Acknowledgement

Writing this part is a rather difficult task. Although the list of people to thank for their contributions is long, making this list is not the hard part. The difficult part is to search for the words that convey the sincerity and magnitude of our gratitude for their contributions.

First, we would like to thank all authors for their contributions. It is their work that has made this book possible. Second, we would like to thank everyone who was involved in the review process. Providing high-quality reviews, which help authors to improve their chapters, is a very difficult task and requires special recognition. Next, our thanks go to the CRC Press team, in particular Ms Erin Harris, who guided us through the book preparation process.

Finally, we would like to thank all those for whom we had much less time than they deserved, because we were editing this book. Thus, our thanks go to our families, colleagues, collaborators and students. Now that the book is ready, we will be with you more often.

Editors

Rashmi Agrawal is a PhD and UGC-NET qualified, with 18-plus years of experience in teaching and research. She is presently working as a Professor in the Department of Computer Applications, Manav Rachna International Institute of Research and Studies, Faridabad. She has authored/co-authored more than 50 research papers, in various peer-reviewed national/international journals and conferences. She has also edited/authored books and chapters with national/international publishers (IGI global, Springer, Elsevier, CRC Press, Apple academic press). She has also obtained two patents in renewable energy. Currently she is guiding PhD scholars in Sentiment Analysis, Educational Data Mining, Internet of Things, Brain Computer Interface, Web Service Architecture and Natural language Processing. She is associated with various professional bodies in different capacities, a Senior Member of IEEE, a Life Member of Computer Society of India, IETA, ACM CSTA and a Senior Member of Science and Engineering Institute (SCIEI).

Marcin Paprzycki is an Associate Professor at the Systems Research Institute, Polish Academy of Sciences. He has an MS from Adam Mickiewicz University in Poznań, Poland, a PhD from Southern Methodist University in Dallas, Texas, and a Doctor of Science from the Bulgarian Academy of Sciences. He is a senior member of IEEE, a senior member of ACM, a Senior Fulbright Lecturer, and an IEEE CS Distinguished Visitor. He has contributed to more than 450 publications and was invited to the program committees of over 500 international conferences. He is on the editorial boards of 15 journals.

Neha Gupta has completed her PhD at Manav Rachna International University, and she has a total of 14-plus years of experience in teaching and research. She is a Life Member of ACM CSTA, Tech Republic and a Professional Member of IEEE. She has authored and co-authored 34 research papers in SCI/SCOPUS/peer reviewed journals (Scopus indexed) and IEEE/IET conference proceedings in the areas of Web Content Mining, Mobile Computing and Cloud Computing. She has published books with publishers such as IGI Global and Pacific Book International and has also authored book chapters with Elsevier, CRC Press and IGI Global USA. Her research interests include ICT in Rural Development, Web Content Mining, Cloud Computing, Data Mining and NoSQL Databases. She is a Technical Programme Committee (TPC) member in various conferences across the globe. She is an active reviewer for the *International Journal of Computer and Information Technology* and in various IEEE Conferences.

Contributors

Geetha Mary A
SCOPE
Vellore Institute of Technology
Tamil Nadu, India

Bibhudendra Acharya
Department of Electronics and
    Communication Engineering
National Institute of Technology
    Rourkela
Rourkela, India

Parul Agarwal
Department of Computer Science
    and Engineering
Jamia Hamdard
New Delhi, India

Ayush Kumar Agrawal
Department of Electronics
    and Communications
    Engineering
National Institute of Technology
    Delhi
Delhi, India

Rangel Arthur
Faculty of Technology
State University of Campinas
Campinas, Brazil

Shaeela Ayesha
Department of Computer Science
Government College University
Faisalabad, Pakistan

Neha Batra
Department of Computer Science
    Engineering
Manav Rachna International
    Institute of Research and
    Studies
Faridabad, India

Sahar Bayoumi
Department of Information
    Technology
CCIS
King Saud University
Riyadh, Saudi Arabia

Rachna Behl
Department of Computer Science
    Engineering
Manav Rachna International
    Institute of Research and
    Studies
Faridabad, India

Manisha Bharti
Department of Electronics and
    Communications Engineering
National Institute of Technology
    Delhi
Delhi, India

Sanjukta Bhattacharya
Department of Information
    Technology
Techno International Newtown
Kolkata, India

Chinmay Chakraborty
Department of Electronics and
    Communication Engineering
Birla Institute of Technology
Jharkhand, India

Sarah Al Dakhil
Department of Information
    Technology
CCIS
King Saud University
Riyadh, Saudi Arabia

Vania V. Estrela
Department of Telecommunications
Fluminense Federal University
Rio de Janeiro, Brazil

Reinaldo Padilha Franca
Faculty of Technology
State University of Campinas
Campinas, Brazil

Kamal Gulati
Amity School of Insurance,
    Banking, and Actuarial
    Science
Amity University
Noida, India

Muhammad Kashif Hanif
Department of Computer Science
Government College University
Faisalabad, Pakistan

Syed Imtiyaz Hassan
Department of Computer
    Science and Information
    Technology
School of Technology
Maulana Azad National Urdu
    University
Hyderabad, India

Yuzo Iano
Faculty of Technology
State University of Campinas
Campinas, Brazil

Shrida Kalamkar
SCOPE
Vellore Institute of Technology
Tamil Nadu, India

Indu Kashyap
Faculty of Engineering and
    Technology
Manav Rachna International
    Institute of Research and Studies
Faridabad, India

K. K. Mahapatra
Department of Electronics and
    Communication Engineering
National Institute of Technology
    Rourkela
Rourkela, India

Setareh Majidian
Faculty of Information Technology
    Management
Allameh Tabataba'i University
Tehran, Iran

Melody Moh
Department of Computer Science
San Jose State University
San Jose, CA

Teng-Sheng Moh
Department of Computer Science
San Jose State University
San Jose, CA

Ana Carolina Borges Monteiro
Faculty of Technology
State University of Campinas
Campinas, Brazil

Ochin Sharma
Department of Computer Science
    Engineering
Manav Rachna International
    Institute of Research and
    Studies
Faridabad, India

Vijay K. Sharma
Department of Electronics
    and Communication
    Engineering
National Institute of Technology
    Rourkela
Rourkela, India

Ramzan Talib
Department of Computer Science
Government College University
Faisalabad, Pakistan

Yihang Tang
Department of Computer Science
San Jose State University
San Jose, CA

Iman Raeesi Vanani
Faculty of Management and
    Accounting
Allameh Tabataba'i University
Tehran, Iran

Section I

Applications of Machine Learning

1

Machine Learning Classifiers

Rachna Behl and Indu Kashyap

CONTENTS

1.1 Introduction

Usually, the term "machine learning" is interchangeable with artificial intelligence, however machine learning is in fact an artificial intelligence sub-area. It is also defined as predictive analysis or predictive modeling. Defined in 1959 by Arthur Samuel, an American computer scientist, the term "machine learning" is the ability of a computer to learn without explicit programming. To predict output values within a satisfactory range, machine learning uses designed algorithms to obtain and interpret input data. They learn and optimise their operations as new data is fed into these algorithms to enhance performance and develop intelligence over time. At present, there are several categories of algorithms for machine learning and they are largely classified as supervised, semi-supervised, unsupervised and reinforcement. Classification is the supervised learning process where classes are sometimes referred to as targets/labels or categories to predict the class of given data points. The machine learning programs draw conclusions in classification from given values and find the category to which new data points pertain. For example, in the context of spam and non-spam classification of emails, the program works on existing data (emails) and filters out the emails as "spam" or "not spam".

Machine learning is a science to use algorithms that help to dig knowledge out of vast amounts of available information (Alpaydin 2014). The Internet of Things, on the other hand, is a buzzword that connects every device using sensors that generate a lot of information. That is the reason machine learning is applied to present-day Internet of Things (IoT) applications. This chapter covers a deep understanding of what machine learning is, how it is related to the IoT and what steps need to be taken while developing a machine-learning-based application.

1.2 Machine Learning Overview

According to Samuel (1959), "machine learning is the ability of computers to learn to function in ways that they were not specifically programmed to do."

Many factors have contributed to making machine learning a reality. These include data sources that generate vast amounts of information, increased computational power for processing that information in fractions of seconds, and algorithms that are now more reliable and efficient. With the IoT in the picture machine learning has grown significantly. Many organisations use IoT platforms for business operations and consulting services. According to (Lantz 2013), with the rapid use of Internet-connected sensory devices, the volume of data being generated and published will increase. As the IoT sensors gather data, an agent analyzes, processes and classifies the data and ensures that information is sent back to the device for improved decision-making. Driverless cars, for example, have to make quick decisions on their own without any human intervention. That's where machine learning comes into play. So, what is machine learning?

It is a data-analysis technique dealing with the development and evaluation of algorithms. It is the science that gives power to computers to act without being explicitly programmed. "It is defined by the ability to choose effective features for pattern recognition, classification, and prediction based on the models derived from existing data" (Tarca et al. 2007).

1.2.1 Steps in Machine Learning

A machine learning task is a collection of various subtasks as shown in the Figure 1.1.

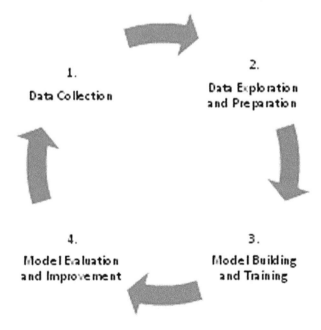

1.
Data Collection

2.
Data Exploration and Preparation

4.
Model Evaluation and Improvement

3.
Model Building and Training

FIGURE 1.1
Steps of machine learning.

TABLE 1.1

Sample Iris Data Set

		FEATURES			
sepal.length	sepal.width	petal.length	petal.width	variety	
5.1	3.5	1.4	0.2	Setosa	
4.9	3	1.4	0.2	Setosa	
4.7	3.2	1.3	0.2	Setosa	
7	3.2	4.7	1.4	Versicolor	INSTANCES
6.4	3.2	4.5	1.5	Versicolor	
6.9	3.1	4.9	1.5	Versicolor	
5.5	2.3	4	1.3	Versicolor	
6.1	3	4.9	1.8	Virginica	
6.4	2.8	5.6	2.1	Virginica	
7.2	3	5.8	1.6	Virginica	

In data collection, data required for analysis is gathered from various sources such as web pages, emails, IoT sensors, text files etc. This data serves as the input to machine learning algorithms, to generate insights from it. The input data may be acquired in any format, but basic the form remains the same, that is, in the form of instances and features. A feature is a trait or attribute that uniquely identifies the sample, and might be useful for learning the desired concept (Lantz 2013). A feature could be numeric, nominal or categorical. A numeric variable is measured in numbers, such as the height of a person or the temperature, whereas a categorical variable is represented by a set of various categories, for example color of eyes: blue, green, brown, black. An example of an iris data set (Fisher 1936) is shown in Table 1.1. There are four features that distinguish one instance from the other. Attribute variety is categorical whereas the other three attributes are numeric in nature. Type of the features help to determine the kind of machine learning algorithm to model.

Any machine learning project is based on the quality of data it uses. The next step, that is, data exploration and preparation, is concerned with a deep study of data so as to prepare high-quality data by cleaning it, removing null values from it, detecting outliers (any suspicious value), removing unwanted features etc. Different techniques of preprocessing are: treating Null Values, Standardisation, Handling Categorical Variables, One-Hot Encoding and Multicollinearity.

The model is then built by training the machine using the algorithms and other machine learning techniques depending on what kind of analysis is required: descriptive, predictive or prescriptive. We may use different approaches like supervised, semi-supervised and unsupervised, or any other approach for developing a model. These approaches are discussed in Section 1.3.

Once the model is built, it is required that it is evaluated and validated for accuracy and other performance measures, like precision, recall etc (Powers 2011). Sometimes a completely different model is built for improving performance.

1.2.2 Performance Measures for Machine Learning Algorithms

Performance measures, in the opinion of Lantz (2013) are used to assess a machine learning algorithm. They provide insight into whether the learner is doing well or not. They also let us to compare two algorithms based on how accurate their results are or how valid the results are. Some commonly used measures are based on creating a confusion matrix.

1.2.2.1 Confusion Matrix

A confusion matrix, as suggested by Lantz, is a table that organises the prediction results of a classification model as:

True Positive (TP): Instance is positive and is labeled as positive

False Positive (FP): Instance is positive and is labeled negative

True Negative (TN): Instance is negative and is labeled negative

False Negative (FN): Instance is negative and is labeled positive

Table 1.2 below displays a confusion matrix used to calculate the predicted class and the actual class.

Based on a confusion matrix the following performance measures can be calculated:

Accuracy, also known as success rate, is formalised as

$$Accuracy = (TP + TN)/(TP + TN + FP + FN)$$

$$Error\,rate = (FP + FN)/(TP + TN + FP + FN)$$

In terms of accuracy error rate $= 1 - accuracy$

TABLE 1.2

Confusion Matrix

	Predicted Class(A)	Predicted Class(B)
Actual Class(A)	TP	FN
Actual Class(B)	FP	TN

Other indicators are precision and recall, which describe how interesting and relevant the model results are, and are calculated as:

Precision is a how often the model is correct in predicting the positive class whereas recall is an indication of completeness (Lantz 2013)

$$Precision = TP/TP + FP$$

$$Recall = TP/TP + FN$$

A model with high precision and high recall is recommended over others. Various other measures such as kappa measure, sensitivity and specificity and f-measure can also be calculated to support the prediction results.

We can use many approaches, such as supervised, semi-supervised or unsupervised, to build a machine learning model. Section 1.3 describes the techniques, highlighting the features of each.

1.3 Machine Learning Approaches

Machine learning is applied in various domains, for example to automate mundane tasks or to have intelligent insight into business. Industries in every sector, whether health, mobile or retail, benefit. You also may be using a device, like a fitness tracker or an intelligent home assistant, that uses machine learning.

In machine learning, computers receive input data, apply statistical learning techniques to automatically identify patterns in data and predict an output. These techniques can be used to make highly accurate predictions (El Naqa and Murphy 2015; Jordan and Mitchell 2015; Sugiyama). We can apply a supervised or an unsupervised approach to generate a model. Recently, semi-supervised and reinforcement learning have been gaining in popularity. The focus of this section is to describe various techniques for model building.

1.4 Types of Machine Learning

Machine learning (ML) is a category of algorithms that allows software applications to be trained and to learned without being explicitly programmed. Machine learning algorithms come in three types based on how learning is performed or how feedback on the learning is provided to the developed system. Figure 1.2 depicts these learning algorithms.

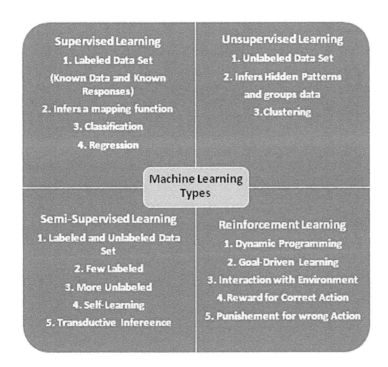

FIGURE 1.2
Types of machine learning.

1.4.1 Supervised Learning

How do you teach a child to distinguish between an animal or a bird? You describe to them the characteristic features of animals, like four legs, a long tail, that they walk etc. and that birds have features like feathers, sharp beaks, fly high in sky etc. If you then show them a picture of a particular creature, they will be able to better distinguish if it is a bird or animal. This is supervised learning performed under the guidance of a teacher. Supervised learning is the most popular paradigm for machine learning, which learns from the labeled data (Kavakiotis, Tsave et al. 2017). A function is inferred from the data that maps the input, to the target, h: $f(x){\rightarrow}y$, f is the function learned from input and output pairs x and y respectively.

There are two further types: classification and regression. Classification predicts categorical response and function learns the class code of different classes that is (0/1) or (yes/no). Naïve Bayes, decision tree, support vector machines (SVM) are commonly used algorithms for classification (Neelamegam and Ramaraj 2013). Regression predicts the quantitative response, for example, predicting the future value of stock price. Linear regression, neural networks, regularisation are algorithms used for regression tasks.

1.4.2 Unsupervised Learning

This is a technique that learns by itself. Compared to supervised learning, no teacher is present. This means the data set is not labeled. It finds the hidden patterns and relationships in the data. For example, if you are given a basket of fruit and you are asked to segregate the fruits, then based on their color, shape, size you can create different clusters of the fruit. Unsupervised learning is data- driven as the basis of it is data and its properties. One category of unsupervised learning is clustering, where the data is organised into groups based on their similarity. K-means clustering, self organising maps, fuzzy c-means clustering are popular techniques under this umbrella.

1.4.3 Semi-Supervised Learning

Supervised Learning works on labeled data, whereas unsupervised technique works on unlabeled data. Practically, getting labeled data is a cumbersome and time-consuming task; we need experts who perform labeling manually, whereas non-labeled data is easily obtained. Semi-supervised learning is a type of learning in which the model is trained in a combination of labeled and unlabeled data. Typically, there is a large amount of unlabeled data compared to labeled. Familiar semi-supervised learning methods are generative models, semi-supervised support vector machines, graph Laplacian based methods, co-training, and multiview learning. These methods form different assumptions on the association between unlabeled data distribution and the classification function. Some of the applications based on this learning approach are speech analysis, protein sequence classification and internet content classifications (Stamp 2017). Recently Google has launched a semi-supervised learning tool called Google Expander.

1.4.4 Reinforcement Learning

Reinforcement learning (RL) combines the dynamic programming and supervised learning approaches to generate efficient machine-learning systems. It is goal-driven learning, where rewards are provided for every correct action and punishment for wrong ones. The machine then learns how to achieve that goal by trial-and-error interactions with its environment. During these interactions actions are performed that may be rewarded or penalised depending on whether they are correct or wrong, such that reward is maximised. This type of learning is the basis for many applications, like game playing, industrial simulations and for resource management applications. Reinforcement learning has a policy, a reward signal, a value function and a model of the environment as its major components (François-Lavet, Henderson et al. 2018). Policy defines a mapping from states to action. A reward signal indicates the award given if a correct step is taken, whereas value function determines the reward in the long run. The model of the environment depicts the behavior

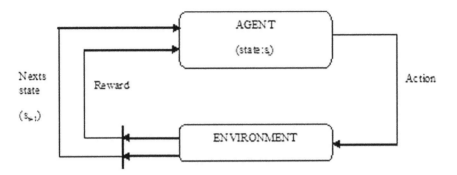

FIGURE 1.3
Reinforcement learning.

of the environment. It is a mapping from state-action to preferred states and rewards. Figure 1.3 shows the interaction of an agent with the environment by taking action and getting reward in return.

Every time the action is taken at state s_t, it is awarded or punished based on the criterion of whether it is correct or not and there is a change in the state from s_t to s_{t+1}.

Machine learning is an interdisciplinary field. Various techniques like supervised, unsupervised, semi-supervised or reinforcement learning can be applied to build the machine learning model, and make a machine learn. The focus of this section is to discuss various machine learning approaches. The next section is related to classification and its different types.

1.5 A Taste of Classification

According to the dictionary, "Classification means the process of deciding into which category an object belongs to based on certain features," for example, categorising a disease based on symptoms. Classification in machine learning is a supervised algorithm and is concerned with building a model that separates data into distinct classes. Classes are also called targets/labels or categories. Input to the model is training samples in which classes are pre-assigned. The model learns the class using this training data and predicts the class of new data. Formally, a classification problem is defined as learning a mapping function $y = f(x,y)$ from input variables x to outcome y depending upon example input–output pairs {x,y}. Output variable y is a categorical variable with different classes. Classification is used to detect if an email is spam or not, to categorise transactions as fraudulent or authorised and many more. There are different forms of classification. The classification types differ from each other based on the kind of values the outcome variable can take, the number of classes being classified and the relationship that exists between independent and dependent variables. We will be discussing each technique in this section.

1.5.1 Binary Classification

Suppose that, looking at the weather conditions, we want to figure out if we should go for an outing or not. In this scenario there are two possibilities: to go or not to go. This is binary classification where only two labels are defined, either 1 or 0. Thus binary classification is a type of supervised learning in which there are only two classes of the outcome variable and the training dataset is labeled. Algorithms applied for binary classification are logistic regression, naïve Bayes, SVM, decision tree.

Some typical applications include:

- Credit card fraudulent transaction detection to classify the transaction as fraud or not.
- Medical diagnosis to define if a disease is cancerous or not.
- Spam detection to identify a mail as spam or ham.

1.5.2 Multiclass Classification

Multiclass or multinomial classification is another variant of a classification problem where N>2 classes are available for outcome variable and each instance belongs to one of three or more classes. The goal is to identify the class that a new data point belongs to by constructing a function. For example, to classify a fruit as banana, apple, orange, various features like shape, color, radius are used. Some other examples are to identify the type of animals in a picture. Various algorithms like KNN, decision trees, SVM can be applied for classifying instances into one of the N classes.

1.5.3 Multilabel Classification

Multilabel classification is applied when there are non-mutually exclusive multiple labels. The target can be assigned to a set of class labels. For example, a news article about politics might also relate to religion or crime or may belong to none of these. It is different from multiclass classification as a sample is assigned only one label.

1.5.4 Linear Classification

A linear classifier classifies the objects into predefined classes based on a linear combination of feature vectors. The boundary of separation between two classes is usually a line or plane. For example, to classify the given objects shown in Figure 1.4, a line is the best separating surface that can be drawn.

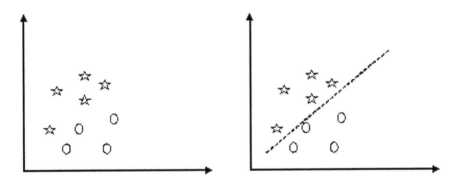

FIGURE 1.4
Linear classification.

Two approaches that can be used for linear classifier are:

- Discriminant function, defined as $y(x) = \sum w_i x_i + b$, where w_i is the weight assigned to each feature vector x_i. Depending upon whether $y>0$ or $y<0$, the class of target variable could be Class 1 or Class 2. If $y = 0$, the target lies on a boundary line. Various discriminant functions can be used like least square, Fischer's method or perceptron.

- Probabilistic approaches determine the class conditional densities for each class. Commonly used methods are logistic regression using sigmoid function and Bayes' theorem.

1.5.5 Non-Linear Classification

Non-linear is a type of classification when data points are not linearly separable. For example, consider the following arrangements of data points shown in Figure 1.5 in which no linear separation exists, so linear methods of classification cannot be applied.

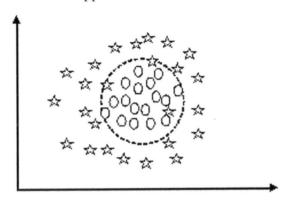

FIGURE 1.5
Non-linear classification.

The best function that can be used to classify these data points is a circle. Other functions, like polynomial kernels, Gaussian kernels, hyperbolic and sigmoid kernels exist for non-linear classification.

In this section we have covered various machine learning classifiers and different scenarios where they can be applied. We have highlighted the key features of each technique and different algorithms that can be used to implement these. In the following section various machine learning classifiers and use of Python to build these is discussed in detail.

1.6 Machine Learning Classifiers

Classification is a supervised technique that predicts the class to which a given data point belongs. A classifier employs input training data to understand the relationship between various features and the outcome (Pagel and Kirshtein 2017). In this section, we will introduce Python programming language as a tool to build machine learning models. Our focus is to describe various classifiers and the use of Python to create the model.

1.6.1 Python for Machine Learning Classification

The Python community has developed many libraries and powerful packages to help researchers implement machine learning. A Python package is a collection of functions that can be shared among users (Müller and Guido 2016). This chapter covers popular machine learning classifiers and their implementation in Python.

We will be using sklearn package to import iris dataset, to build various classifiers and to compute their score. sklearn or scikit-learn (Pedregosa, Varoquaux et al. 2011) is a powerful open source machine learning library available in Python. It provides a wide variety of supervised algorithms like decision tree, random forest, SVM, k-nearest neighbor.

The first step in developing any machine learning model using Python is to import the necessary libraries.

```
# import required libraries
from sklearn import datasets
from sklearn.metrics import confusion_matrix
from sklearn.model_selection import train_test_split
import numpy as np
import matplotlib.pyplot as plt
import matplotlib.image as mpimg
import pandas as pd
```

The next step is to load the dataset, divide it into train and test set

```
# load iris dataset
iris = datasets.load_iris()

# Split into features and label
X = iris.data
y = iris.target

# Divide the dataset as train and test data
X_train, X_test, y_train, y_test = train_test_split(X,
y, random_state = 0)
```

The next step is to build the model and evaluate its performance. We will be discussing building different classifiers in the respective classifier section.

1.6.2 Decision Tree

Decision tree is one of the most popular supervised learning algorithms. It was introduced by Leo Breiman in 1984 in his book *Classification and Regression Trees*. It can be used for both classification and regression problems. When applied for classification tasks, a decision tree is called a classification tree, whereas it is known as a regression tree if it is employed for regression. We will focus mainly on decision trees as classification trees.

Decision tree uses a tree-like structure with one topmost node as the root node and internal nodes and branches (Witten, Frank et al. 2016). An internal node is that characteristic (or attribute) on which the classification task will be based, the branch represents a decision rule (a test condition) and each leaf node represents the result (one out of predefined set of classes). Initially, there is a single node in a decision tree that branches into various results. Each of the resulting nodes further branch off into other possible outcomes resulting in a tree-like structure. For instance, if we want to judge whether someone likes computer games or not given his or her age, gender, occupation etc. the decision tree as depicted in Figure 1.6 can be generated.

Once a decision tree is created, classification rules are derived out of it. Classification rule is an if–then–else rule that considers all the scenarios and assigns class variables to each. The following classification rules are generated from the decision tree created:

R1: IF (GENDER='M') AND (OCCUPIED='Y') THEN Play Games='Yes'

R2: IF (GENDER='M') AND (OCCUPIED='N') and (AGE<15) THEN Play Games='No'

R3: IF (GENDER='M') AND (OCCUPIED='N') AND (AGE>15) THEN Play Games='Yes'

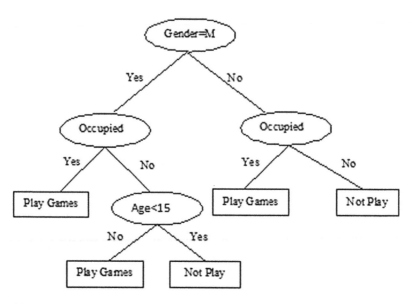

FIGURE 1.6
Decision tree.

R5: If (GENDER='F') AND (OCCUPIED='Y') THEN Play Games='Yes'
R6: IF (GENDER='F') AND (OCCUPIED='N') THEN Play Games='No'

1.6.2.1 Building a Decision Tree

Decision trees are type of learning algorithms that use information as main source of learning. The goal of decision trees is to find those informative features that have most "information" about the outcome variable. Building a decision tree model is a two-step process. The first process is induction where the tree is built by learning decision rules based on information in the data. While constructing a decision tree, it may be subjected to overfitting by the learning process applied. Pruning helps to remove unimportant structures from the decision tree, thereby reducing the complexity. Let us elaborate the induction step to build a decision tree.

1.6.2.2 Induction

The induction process consists of recursive steps that use a greedy approach to make an optimal local choice at each node. It is a top-down approach of creating a decision tree, and various decision trees inducers such as ID3 (Quinlan 1986), C4.5 (Quinlan 1993), and CART (Breiman et al. 1984; Wu, Kumar et al. 2008), exist in machine learning. Some have two phases: growing and pruning as in C4.5 and CART. Other inducers have only the growing phase.

1. Initially a whole training set having feature variables and outcome variables is considered as the root node.
2. Determine the "best feature" in the dataset to split the data on. Finding the best feature depends on the splitting index. Various schemes have been proposed but we will discuss later two main schemes, like Entropy and Gini Index.
3. Split the training sets into subsets such that each subset has possible values for the best attribute. This creates a node on the tree.
4. Repeat Steps 2 to 3 to generate new tree nodes by using the subset of data. Stop when all leaf nodes are determined or when some measures like accuracy and number of nodes/splits is optimised.

1.6.2.3 Best Attribute Selection

Entropy is the measure of homogeneity. A homogeneous data set has 0 entropy and if it is divided, entropy is 1. Consider the data set of Table 1.3 for tree creation. We can create as many trees as there are attributes. But which tree is optimal is decided by which best attribute is the root. Let us illustrate this by taking an example data set as shown in Table 1.3. We want to classify whether the user buys a computer based on his or her age, income, credit-rating, and whether he or she is a student or not.

The first step is to find the entropy of target variable which is Buys_Computer. Entropy (S) for a data set S containing m classes are defined as:

$$\text{Entropy}(S) = -\sum_{j=1}^{m} Pj LogPj \qquad (1.1)$$

TABLE 1.3

Specimen Data Set

S.No	Age	Income	Student	Credit_Rating	Buys_Computer
1	<=30	High	No	Excellent	No
2	<=30	High	No	Fair	No
3	31-40	High	No	Fair	Yes
4	>40	Medium	No	Fair	Yes
5	>40	Low	Yes	Fair	Yes
6	>40	Low	Yes	Excellent	No
7	31-40	Low	Yes	Excellent	Yes
8	<=30	Medium	No	Fair	No
9	<=30	Low	Yes	Fair	Yes
10	31-40	Medium	Yes	Fair	Yes
11	<=30	Medium	Yes	Excellent	Yes
12	31-40	Medium	No	Excellent	Yes
13	31-40	High	Yes	Fair	Yes
14	>40	Medium	No	Excellent	No

$$P(\text{Buys\_Computer} = \text{Yes}) = 9/14 \qquad (1.2)$$

$$P(\text{Buys\_Computer} = \text{No}) = 5/14 \qquad (1.3)$$

$$\text{Entropy}(\text{Buys\_Computer}) = -9/14\log(9/14) - 5/14\log(5/14) = 0.940 \qquad (1.4)$$

Next find Information Gain of each attribute using:

$$\text{Gain}(S, A) = \text{Entropy}(S) - \sum_{v \in A} \left(\frac{|Sv|}{|S|} * \text{Entropy}(Sv) \right) \qquad (1.5)$$

Now Entropy(Age) is

$$E(age) = \frac{5}{14} I(2,3) + \frac{4}{14} I(4,0)$$
$$+ \frac{5}{14} I(3,2) = 0.694 \qquad (1.6)$$

$$\text{So Information Gain}(\text{Age}) = 0.940 - 0.694 = 0.246 \qquad (1.7)$$

Similarly,

$$\text{Information Gain}(\text{Income}) = 0.029 \qquad (1.8)$$

$$\text{Information Gain}(\text{Student}) = 0.151 \qquad (1.9)$$

$$\text{Information Gain}(\text{Credit\_Rating}) = 0.048 \qquad (1.10)$$

As Information Gain for Age is the highest of all the attributes, it is thus the best attribute for splitting. Information Gain of various attributes is compared at each split to decide which among the attributes is to be chosen for splitting.

Thus, the optimal tree created after selecting age as a splitting attribute looks like Figure 1.7.

Another measure to find the optimal attribute is Gini Index and can be calculated as:

$$\text{Gini}(S) = 1 - \sum_{j=1}^{m} Pj^2 \qquad (1.11)$$

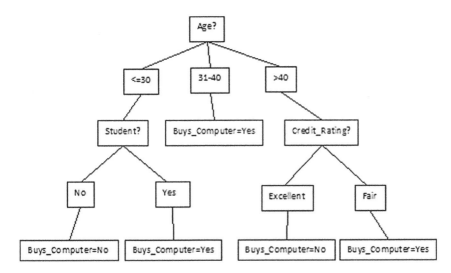

FIGURE 1.7
Optimised decision tree.

where Pj is the relative frequency of class j in S. Using the entropy or the gini index, information gain can be computed as:

$$Gain(S,A) = Gini(S) - \sum_{v \in A} \left(\frac{|Sv|}{|S|} * Gini(Sv) \right) \qquad (1.12)$$

where v represents any possible values of attribute A; Sv is the subset of S for which attribute A has value v; |Sv| is the number of elements in Sv; |S| is the number of elements in S.

1.6.2.4 Pruning

In machine learning and data mining, pruning is a technique to reduce the size of decision trees by eliminating those tree structures that are not effective in classification. Decision trees are more prone to overfitting and pruning can reduce it. Two approaches to prune decision trees are:

- Pre-pruning: This technique stops the growth of the tree construction prematurely by evaluating measures such as information gain, gini index etc.
- Post-pruning: allows the tree to grow completely, and then starts pruning the tree.

The code below specifies how to build a decision tree classifier and create a confusion matrix.

```
# create DescisionTree model
from sklearn.tree import DecisionTreeClassifier
dt_model = DecisionTreeClassifier(max_depth = 2)
dt_model.fit(X_train, y_train)
dt_predictions = dt_model.predict(X_test)

# create confusion matrix
cm = confusion_matrix(y_test, dt_predictions)
dt_model.score(X_test,y_test)

# output
0.8947368421052632
```

Decision tree is simple and easy to understand and it can be easily visualised. It can handle both quantitative and qualitative data. It is a kind of white-box strategy of machine learning algorithm. However, it suffers from the problem of overfitting (a condition when the model is trained with so much data that it learns the noise and fluctuations in the sample training data). Decision tree is susceptible to variation in data, that is, it is unstable and small changes in data can generate a totally different tree. Also generating an optimised decision tree is NP-complete.

1.6.3 Random Forests

A random forest is a supervised learning algorithm that creates a forest consisting of multiple random decision trees. A randomised variant of a tree induction algorithm is applied to build the group (or forest) of decision trees. The randomness can be of two types: the tree can be built by selecting random samples from the original data or a subset of features is selected at random to generate the best split at each node. But why we are going to use a random forest? What are the benefits? Random forest is a family of supervised learning and is generally used for classification. Unlike decision tree, random forests are not prone to "overfitting." Using multiple trees reduces the risk of overfitting. Their training time is also less. They produce highly accurate predictions and can be run efficiently on large database. Accuracy is maintained even if a large proportion of the data is missing. Some of the application domains of random forest are remote sensing and multiclass object detection. The algorithm is also used in the game console "Kinect." In the next section we will see how random forests are generated, trained and interpreted.

A random forest works in the following way:

- In contrast to decision tree, where the whole data set is considered, random samples are created using bagging (bootstrap aggregating). A new dataset is created containing m out of n cases, selected at random with replacement. Some rows that are left out are known as out-of-bag-samples (OOB) and they account for around one third of the total samples.

- First, decision tree is created using the bootstrap sample. For this, a random subset of variables/attributes is used. Out of these variables the best significant predictor is selected as the root node. The process is repeated for each of further branch nodes, that is, randomly selecting variables as candidates for the branch node and then choosing the variable that best classifies the samples. Selecting the optimal number of variables is a research question. Usually, it is p/3 for regression tree and $\sqrt{p}$ for classification trees.

- Several trees are grown by repeating the previous steps. How many trees to build is again a research question. Each decision tree predicts the output class based on the predictor variables that are used in tree creation. The final prediction is obtained by averaging or voting.

1.6.3.1 Evaluating Random Forest

Random forest is evaluated for accuracy by using out-of-bag samples. The accuracy of random forest can be measured by the percentage of OOB samples that are correctly classified and those OOB samples that are misclassified form the out-of-bag error.

1.6.3.2 Tuning Parameters in Random Forest

While building the random forest, low correlation and reasonable strength of the trees is desired. Breiman (2001) suggested various metrics such as mtry, sample size and node size to control this.

1.6.3.2.1 mtry

mtry is the number of random variables that are selected for building the tree. A lower value of mtry generates more different, less related trees that have better constancy when aggregating. Low mtry is important as its higher values might mask the less important features with strong attributes. On the other hand, the lower values of mtry generate trees built on suboptimal variables leading to worse average performance. The trade-off between stability and the accuracy of the individual tree has to be dealt with. Bernard et al. (2009) conclude that "mtry $=\sqrt{p}$ is a reasonable value, but can sometimes be improved. If there are many predictor variables, mtry should be set low." However, it should be set to a high value if the predictors are few. Genuer et al. (2008) also suggested that "fixing mtry as $\sqrt{p}$ as it is convenient regarding the error rate." "Computation time decreases approximately linearly with lower mtry values, since most of RF's computing time is devoted to the selection of the split variables." (Wright and Ziegler 2017).

1.6.3.2.2 Sample Size

Another parameter that affects the performance of random forest is sample size. Martínez-Muñoz and Suárez (2010) carried out a first-hand analysis to discover the relationship between performance of the tree and sample size. They came to conclusion that the

> optimal value is problem dependent and in most cases performance is better if less observations are selected as sample than the standard choice (which is to sample as many observations with replacement as the number of observations in the dataset). Setting it to lower values reduces the runtime.

The authors also claimed no performance difference is observed if sampling is done with or without replacement.

1.6.3.2.3 Number of Trees

This parameter, although it may not be tunable, should be set high as suggested by Probst and Boulesteix (2017). Usually, the value depends on the data set properties and the required convergence rate. Performance is gained on the first 100 trees grown. As the trees are trained independently, many software packages, such as ranger, provide parallel learning on several available CPU cores.

1.6.3.3 Splitting Rule

The splitting rule decides the candidate that is best out of the randomly selected variables at each node. Many splitting rules have been defined by researchers (Hothorn et al. 2006; Breiman 2001; Hothorn and Zeileis 2015). A more commonly used rule is defined by Breiman and is based on the selection of variables based on Gini index. The split that minimises the Gini index value is the best. Another splitting rule is based on the p-value of the global test. Although a lot of work is carried out in this area the best split is often decided by the data set properties.

Python Code for Random Forest

```
from sklearn.ensemble import RandomForestClassifier

#Build Random forest model using Gaussian Classifier
clf=RandomForestClassifier(n_estimators=100)

#Train the model using the training sets
clf.fit(X_train,y_train)
y_pred=clf.predict(X_test)

#use scikit-learn metrics library for finding accuracy
from sklearn import metrics
# print accuracy
print("Accuracy:",metrics.accuracy_score(y_test, y_pred))
```

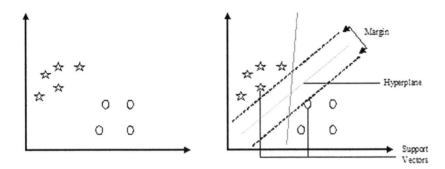

FIGURE 1.8
Support vector machine.

1.6.4 Support Vector Machine

Support vector machines developed by Cortes and Vapnik (1995) are used for binary classification. It is one of the long-established classical machine learning techniques that can solve classification problems. SVM is a supervised learning technique for classification and regression. "Given a set of p-dimensional vectors in vector space, SVM finds the separating hyperplane that splits vector space into sub-set of vectors; each separated subset (so-called data set) is assigned by one class" (Nguyen 2017). Let us understand the concepts of support vectors using an example shown in Figure 1.8.

Initially an arbitrary hyperplane is drawn and its distance is measured from the closest data points known as support vectors. The distance between hyperplanes and support vectors is known as the margin. Many separating hyperplanes can be drawn but the one that maximises the margin between two subsets is the optimal one. As the dataset is linearly separable, the hyperplane defined is a linear line and is represented as H:$w^t(X)$ + B, where w is a normal vector to the hyperplane. A data point with $w^t(x)$ + B>0 will belong to one class and the point where $w^t(x)$ + B<0 will belong to another class in two-class problems. If $w^t(x)$ + B = 0, the point will lie on the hyperplane. In the figure, two hyperplanes are shown describing the above concepts

Sometimes, however, the points are not linearly separable, for example as shown in Figure 1.9.

In this scenario kernels are used to map non-linear input data to high-dimensional feature space. The new mapping is then linearly separable. Examples of kernels are Polynomial Kernel, Gaussian Kernel, Sigmoid Kernel and many more.

To build a svm classifier the following Python code can be used.

```
# fit SVM classifier using linear kernel
from sklearn.svm import SVC
```

```
svmmodel = SVC(kernel = "linear", C = 1)
svmmodel.fit(X_train, y_train)
svm_prediction = svmmodel.predict(X_test)

# calculating model accuracy
accuracy = svmmodel.score(X_test, y_test)

# creating a confusion matrix
cm = confusion_matrix(y_test, svm_predictions)
svmmodel.score(X_test,y_test)

#output
0.9736842105263158
```

1.6.5 Neural Networks

A neural network belongs to a class of naturally inspired algorithms (Shanmuganathan 2016). It is based on the biological system found in the human brain. Like the human brain, a neural network consists of semantic networks of neurons (input/output units) connected to each other. The connections are generally weighted to show the importance of the link between one neuron and the next. The neuron is the basic computation unit of a neural network. These neurons are interconnected in large numbers and are

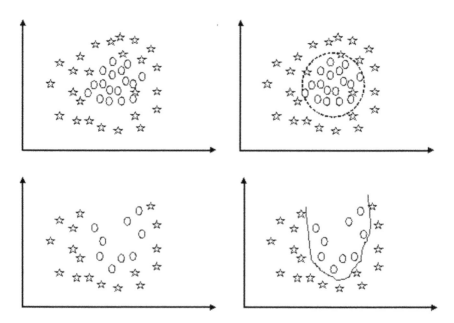

FIGURE 1.9
Non-linear classification using SVM.

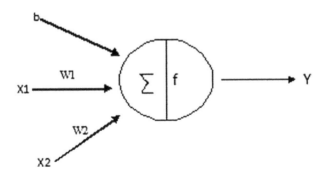

FIGURE 1.10
Basic neuron.

stacked into multiple layers. The structure of a single neuron is also called a perceptron, as shown in Figure 1.10.

In the Figure x1 and x2 are the inputs to the neuron/node. w1 and w2 are weights assigned to the inputs. x1 and x2 describe their relative importance. Output is generated by applying activation function f on the weighted sum of all the inputs.

$$Y = w1 \times 1 + w2 \times 2$$

$$\text{ie } Y = f\left(\sum x_i w_i\right)$$

Various activation functions can be used depending on the type of classification problem being addressed. Commonly used activation functions are linear, step, tanh, sigmoid. Many more can be used, as shown in Table 1.4.

Figure 1.11 displays a multilayer neural network containing interconnected neurons stacked into different layers. The layers of neural network are:

Input layer
Hidden layer
Output layer

There is one input layer consisting of multiple neurons, one hidden layer and one output layer. This is an example of a multilayer network. The number of hidden layers can vary from 1 to N (deep neural networks). If the information travels in a forward direction from input layer to hidden layer and then to output layer, this is an example of a feed-forward network.

Training of a neural network is done using the most popular back-propagation algorithm.

TABLE 1.4

Activation Functions

Name	Equation	Derivative
Identity	f(x)=x	$f'(x)=1$
Binary step	$f(x)=\begin{cases} 0 \ for \ x < 0 \\ 1 \ for \ x \geq 0 \end{cases}$	$f'(x)=\begin{cases} 0 \ for \ x \neq 0 \\ ? \ for \ x = 0 \end{cases}$
Logistic	$f(x)=\dfrac{1}{1+e^{-x}}$	$f'(x)=f(x)(1-f(x))$
Tanh	$f(x)=\tanh(x)=\dfrac{2}{1+e^{-2x}}-1$	$f'(x)=1-f(x)^2$
ArcTan	$f(x)=tan^{-1}(x)$	$f'(x)=\dfrac{1}{x^2+1}$
Rectified Linear Unit(ReLU)	$f(x)=\begin{cases} 0 \ for \ x < 0 \\ x \ for \ x \geq 0 \end{cases}$	$f'(x)=\begin{cases} 0 \ for \ x < 0 \\ 1 \ for \ x \geq 0 \end{cases}$
Parametric Rectified Linear Unit(PReLU)	$f(x)=\begin{cases} ax \ for \ x < 0 \\ x \ for \ x \geq 0 \end{cases}$	$f'(x)=\begin{cases} \alpha \ for \ x < 0 \\ 1 \ for \ x \geq 0 \end{cases}$
Exponential Linear Unit(ReLU)	$f(x)=\begin{cases} a\left(e^{-x}-1\right) for \ x < 0 \\ \quad x \ for \ x \geq 0 \end{cases}$	$f'(x)=\begin{cases} f(x)+\alpha \ for \ x < 0 \\ \quad 1 \ for \ x \geq 0 \end{cases}$
Soft Plus	$f(x)=\log_e\left(1+e^x\right)$	$f'(x)=\dfrac{1}{1+e^{-x}}$

1.6.5.1 Back Propagation Algorithm

1. Initially weights are assigned randomly. Output is calculated based on the assigned weights.
2. Compute the difference between actual and observed output.
3. If the difference is large, adjust the weights using delta rule or gradient descent.
4. Repeat step 2 to 3 until weights are found that minimise the error function.

Neural network implementation using Python

```
# Replace target with 0, 1, 2
from sklearn.preprocessing import LabelEncoder
```

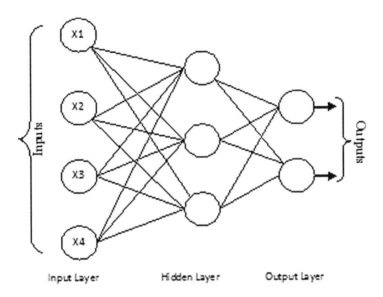

FIGURE 1.11
Multi-layer neural network.

```
lencoder = LabelEncoder()
data["Species"] = l.fit_transform(data["Species"])
species = pd.DataFrame({'Species': ['Iris-setosa',
'Iris-versicolor', 'Iris-virginica']})

# Features before mean normalisation
unscaled_features = X_train

# Mean Normalisation to have a faster classifier
from sklearn.preprocessing import StandardScaler
sc = StandardScaler()

#calculate mean and standard deviation
X_train_array = sc.fit_transform(X_train.values)

# Assign the scaled data to a DataFrame & use the index
and columns arguments to keep your original indices and
column names:
X_train = pd.DataFrame(X_train_array, index=X_train.in
dex, columns=X_train.columns)

# Center test data with the μ & σ computed (fitted) on
training data
X_test_array = sc.transform(X_test.values)
```

```
X_test = pd.DataFrame(X_test_array, index=X_test.index,
columns=X_test.columns)

# import the model
from sklearn.neural_network import MLPClassifier

# Initializing the multilayer perceptron
mlp = MLPClassifier(hidden_layer_sizes(5), solver='sgd',
learning_rate_init= 0.01, max_iter=500)
mlp.fit(X_train, y_train)
print(mlp.score(X_test,y_test))
```

"Neural networks are used in wide range of sectors like medicine, marketing, financial due to their fault tolerance and ability to deal with noisy data but training time required is large" (Cilimkovic 2015).

1.6.6 Logistic Regression

Logistic regression, or logistic model or logit model, examines the relationship between a set of predictor variables and a categorical response variable, and determines the probability of occurrence of an event by modeling the response in terms of predictors using a logistic or sigmoid curve (DeGregory, Kuiper et al. 2018). Logistic regression models are binary logistic regression and multinomial logistic regression depending on whether the dependent variable is binary or not. If the dependent variable is binary, having two values, true or false, and independent variables are either continuous or categorical, binary logistic regression is applied. Multinomial logistic regression is applied when the response variable has more than two categorical values. The relationship between independent and dependent variables is represented as:

$$Y = b0 + b1X1 + b2X2 + \ldots\ldots bnXn \tag{1.13}$$

Where Y is response variable and X1, X2,........Xn are predictor variables. The coefficients b0, b1,b2,.........bn are learned during training.

Logistic regression estimates the probability of occurrence of an event using a logistic function and the standard sigmoid function is defined as:

$$f(x) = \frac{1}{1 + e^{-x}} \tag{1.14}$$

So

$$p(Y) = f(x) = \frac{1}{1 + e^{-b0 + b1X1 + b2x2 + \ldots bnxn}} \tag{1.15}$$

Based on the probability so obtained, the prediction result for binary logistic regression is defined as:

```
0 if p(Y)<0.5
1 if p(Y)>0.5
```

```python
# Import the iris dataset
dataset = pd.read_csv('iris.csv')

#Split the dataset in predictor and reponse variables
X = dataset.iloc[:,:4].values
y = dataset['species'].values

# Feature Scaling to bring the variable in a single scale
from sklearn.preprocessing import StandardScaler
stdscalar = StandardScaler() #initialize StandardScaler
object
X_train = stdscalar.fit_transform(X_train)
X_test = stdscalar.transform(X_test)
# Fit Multiclass Logistic Classification to the Training set
from sklearn.linear_model import LogisticRegression
logisticregression = LogisticRegression()
logisticregression.fit(X_train, y_train)

LogisticRegression(C=1.0, class_weight=None, dual=False,
fit_intercept=True,
    intercept_scaling=1, max_iter=100, multi_
    class='ovr', n_jobs=1,
    penalty='l2', random_state=None, solver='liblinear',
    tol=0.0001,
    verbose=0, warm_start=False)

# Predict the labels
y_pred = logisticregression.predict(X_test)

# Confusion Matrix
from sklearn.metrics import confusion_matrix
cm = confusion_matrix(y_test, y_pred)

#finding accuracy from the confusion matrix.
b= cm.shape
corrPred = 0
falsePred = 0

for row in range(b[0]):
```

```
for c in range(b[1]):
  if row == c:
     corrPred +=cm[row,c]
  else:
     falsePred += cm[row,c]

print('True predictions: ', corrPred)
print('False predictions', falsePred)
print ('Accuracy of the multiclass logistic
classification is: ', corrPred/(cm.sum()))
Correct predictions: 26
False predictions 4
Accuracy of the multiclass logistic regression is:
0.866666666667
```

1.6.7 k-Nearest Neighbor

The k-nearest neighbor (k-NN) algorithm is a non-parametric, instance-based learning technique. It can be used for both classification and regression tasks. As it is non-parametric, it does not make assumptions about the data distribution. In k-NN, algorithm data used for training is labeled with various classes. It stores all the available samples and uses them to classify the newer samples. It does this by finding distance between new instances and available labeled instances using popular distance measuring techniques like Euclidean distance, Manhattan distance, Minkowski. The k-NN algorithm identifies the k-similar training samples where the parameter k is already set. The class of the majority of k-neighbors is assigned to the new instance.

In a k-NN algorithm, each instance is described by n number of features. These instances are considered as points within an n-dimensional feature space. If we plot the instances in the n-dimensional feature space, the similar instances will be plotted near to each other. Their exact position in the feature space is not as important as their relative position to each other. Figure 1.12 shows the scatter plot of data objects.

1.6.7.1 The k-NN Algorithm

Steps to perform classification/regression using kNN are:

1. Load the labeled training data.
2. Set k to specify the number of nearest neighbors.
3. Repeat for every record in the training data:
   a. Compute its distance with the test record using distance metrics.
4. Arrange the distances and record in in ascending order according to the distances.

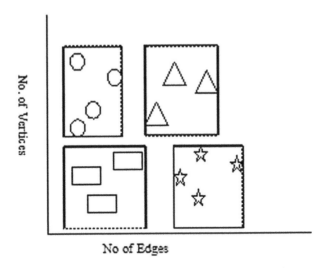

No of Edges

FIGURE 1.12
Scatter plot.

5. Select the top k neighbors' and their labels.

6. Compute the mean of the selected k labels for regression task, and

7. mode for classification task.

from sklearn.neighbors import KNeighborsClassifier

```
import numpy as np
kn = KNeighborsClassifier(n_neighbors=1)
kn.fit(X_train, y_train)
x_new = np.array([[7, 4.3, 3, 0.5]])
prediction = kn.predict(x_new)
print("Predicted target value: {}\n".format(prediction))
print("Predicted feature name: {}\n".format
    (iris_dataset["target_names"][prediction]))
print("Test score: {:.2f}".format(kn.score(X_test, y_test)))
```

1.7 Model Selection and Validation

Machine learning has an important place in contemporary scenarios. Whether these are consumers, customers or businesses, all rely on it to improve decision-making. We are applying predictive modeling techniques to our research or business problems and want to make "good" and effective predictions! Once a model is fit, what is its performance? How well does it respond to unseen data? What is the certainty that this is the best model to be selected for the problem? Keeping in mind these questions, we will now discuss model selection and validation.

Model selection is the process of selecting an approach for the business problem under consideration. It also means choosing various hyperparameters or sets of features for a machine learning approach. A plethora of different machine learning approaches exist, for example, SVM, logistic regression, clustering etc. But which algorithm or machine learning model do we choose from amongst the many available? What are the criteria for choosing a model? A model that is chosen to solve any problem must be simple, interpretable, fast, scalable and accurate. Not only this, model evaluation also has to be considered and investigated for the selected model. Model evaluation aims at estimating the performance of a model on unseen data (Raschka 2018). Various factors that help in the selection of a model are as follows.

1.7.1 Hyperparameter Tuning and Model Selection

Parameters that are specified before model fitting are those of the learning method, and their value is set before learning. For example, the value of k in k-nearest neighbors, is the learning rate of a neural network. To build a good model, the best hyperparameters should be aimed for. To tune hyperparameters, a large dataset is divided up as training set, validation set, test set. As many models are trained using a training set as are the various blends of model hyperparameters. Thereafter, the models are assessed on the validation set, and the one having optimum performance on the validation set is selected. The selected model is again subject to learning on training and validation data and evaluated using the test set. If the generalisation error is the same as the validation error, the model will execute effectively on new instances also. Lastly, the model is retrained on the complete data (train, validation and test set) before it is used in "production." Smaller data sets however are split into training set and test set. Estimates of generalisation error are made on the test set.

1.7.2 Bias, Variance and Model Selection

Bias refers to a phenomenon that occurs due to erroneous assumptions in the learning algorithm whereas variance means how sensitive the algorithm is to small variations in training data. High bias means the model is too simple to capture all the underlying information of the data and thus is generating an under-fit model (Varma and Simon 2006). A model with high variance is prone to noise, and tries to fit most of the data points by making the model complex. High variance causes over-fitting of data. Figure 1.13 explains the relation between bias-variance and model complexity.

So, given two models that equally explain the outcome, the simpler model is a better choice (this is also referred to as Occam's razor as applied in machine

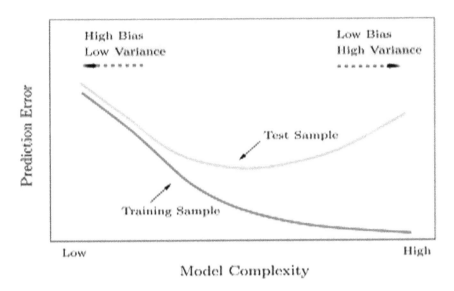

FIGURE 1.13
Basic variance trade off. (Source: Friedman, Hastie and Tibshirani 2001.)

learning). Simpler models with fewer variables are desirable because they lead to lower variance and are easier to interpret.

1.7.3 Model Validation

Once the model is trained and results are acquired, we want to assure the model performance in terms of accuracy and validate the results. Cross validation (CV) is one of the techniques used to test the effectiveness of a machine learning model. It is used to test models on new unseen data and assess whether the model overfits, underfits or works well (Hawkins, Basak, and Mills 2003; Kim 2009).

There are many ways in which cross validation can be performed. These techniques are discussed.

The holdout method is the simplest technique of cross validation. The data set is split into two sets – the training set and the testing set. Generally, the 70% to 30% rule is followed. The model learning is done using the training set only. The learned model is then executed over the test set to predict the output. The mean absolute test set error is calculated to find out how far the predicted output is from the actual output so as to evaluate the model. This method results in high bias when we have limited data because we might skip some important information of the data. Nevertheless, for large data sets, this technique works fairly well.

K-fold cross validation is an improvement over the holdout method. In this technique, the data set is divided into k-testing subsets. Each time, one out of the k subsets is used as the test set and the model is trained using the remaining k–1 subsets. Then the average error across all k trials is computed. As in this technique, every data point gets a fair chance to be in a test set precisely once, and in the training set k–1 times, bias is improved. The disadvantage of the technique is that the training procedure runs for k times.

Leave-one-out cross validation is a special type of K-fold cross validation, with K equal to N, the number of data points in the set. Every time the model is trained using an N–1 sample with one sample held for the test set. Test error is computed on the held sample. After complete training the test error is averaged out. This is a comparatively expensive technique of cross validation.

Conclusion

This book chapter focuses on machine learning classifiers. There are various ways through which machine learning classifiers are built. The authors have tried to differentiate between various learning approaches using examples. Famous machine learning classifiers are presented in the chapter, and the ways to build them using Python programming language are also demonstrated. The chapter also covers the performance measures used to judge the effectiveness of the classifier and techniques to validate the built models.

References

Alpaydin, E., 2014, *Introduction to machine learning*, MIT Press, Cambridge, MA.

Bernard, S., Heutte, L. and Adam, S., 2009, June, 'Influence of hyperparameters on random forest accuracy', in *International workshop on multiple classifier systems*, Springer, Berlin, Heidelberg, pp. 171–180.

Breiman, L., Friedman, J., Olshen, R. and Stone, C., 1984, 'Classification and regression trees. Wadsworth Int', *Group*, vol. 37, no. 15, pp. 237–251.

Breiman, L., 2001, 'Random forests', *Machine Learning*, vol. 45, no. 1, pp. 5–32.

Cilimkovic, M., 2015, *Neural networks and back propagation algorithm*, Institute of Technology, Blanchardstown, Dublin, 15.

Cortes, C. and Vapnik, V., 1995, 'Support-vector networks', *Machine Learning*, vol. 20, no. 3, pp. 273–297.

DeGregory, K. W., Kuiper, P., DeSilvio, T., Pleuss, J. D., Miller, R., Roginski, J. W., Fisher, C. B., Harness, D., Viswanath, S., Heymsfield, S. B., Dungan, I. and Thomas, D. M., 2018, 'A review of machine learning in obesity', *Obesity Reviews*, vol. 19, no. 5, pp. 668–685.

El Naqa, I. and Murphy, M. J., 2015, 'What is machine learning?', in *Machine learning in radiation oncology*, Springer, Cham, pp. 3–11.

Fayyad, U., Piatetsky-Shapiro, G. and Smyth, P., 1996, 'From data mining to knowledge discovery in databases', *AI Magazine*, vol. 17, no. 3, pp. 37–37.

Fisher, R. A., 1936, 'The use of multiple measurements in taxonomic problems', *Annals of Eugenics*, vol. 7, no. 2, pp. 179–188.

François-Lavet, V., Henderson, P., Islam, R., Bellemare, M. G. and Pineau, J., 2018, 'An introduction to deep reinforcement learning', *Foundations and Trends® in Machine Learning*, vol. 11, no. 3–4, pp. 219–354.

Friedman, J., Hastie, T. and Tibshirani, R., 2001, *The elements of statistical learning*, vol. 1, no. 10, Springer Series in Statistics, New York.

Genuer, R., Poggi, J. M. and Tuleau, C., 2008, 'Random forests: Some methodological insights', *arXiv preprint arXiv:0811.3619*.

Hawkins, D. M., Basak, S. C. and Mills, D., 2003, 'Assessing model fit by cross-validation', *Journal of Chemical Information and Computer Sciences*, vol. 43, no. 2, pp. 579–586.

Hothorn, T., Hornik, K. and Zeileis, A., 2006, 'Unbiased recursive partitioning: a conditional inference framework', *Journal of Computational and Graphical Statistics*, vol. 15, no. 3, pp. 651–674.

Hothorn, T. and Zeileis, A., 2015, 'partykit: a modular toolkit for recursive partytioning in R', *The Journal of Machine Learning Research*, vol. 16, no. 1, pp. 3905–3909.

Jiawei, H. and Micheline, K., 2000, *Data mining: concepts and techniques*, Elsevier Book Series.

Jordan, M. I. and Mitchell, T. M., 2015, 'Machine learning: trends, perspectives, and prospects', *Science*, vol. 349, no. 6245, pp. 255–260.

Kavakiotis, I., Tsave, O., Salifoglou, A., Maglaveras, N., Vlahavas, I. and Chouvarda, I., 2017, 'Machine learning and data mining methods in diabetes research', *Computational and Structural Biotechnology Journal*, vol. 15, pp. 104–116.

Kim, J. H., 2009, 'Estimating classification error rate: repeated cross-validation, repeated hold-out and bootstrap', *Computational Statistics and Data Analysis*, vol. 53, no. 11, pp. 3735–3745.

Kohavi, R., 1995, August, 'A study of cross-validation and bootstrap for accuracy estimation and model selection', *Ijcai*, vol. 14, no. 2, pp. 1137–1145.

Kourou, K., Exarchos, T. P., Exarchos, K. P., Karamouzis, M. V. and Fotiadis, D. I., 2015, 'Machine learning applications in cancer prognosis and prediction', *Computational and Structural Biotechnology Journal*, vol. 13, pp. 8–17.

Kubat, M., 2017, *An introduction to machine learning*, vol. 2, Springer International Publishing, Cham, Switzerland.

Lantz, B., 2013, *Machine learning with R*, Packt Publishing Ltd, Birmingham, UK.

Lison, P., 2015, 'An introduction to machine learning', *Language Technology Group (LTG)*, vol. 1, p. 35.

Mahdavinejad, M. S., Rezvan, M., Barekatain, M., Adibi, P., Barnaghi, P. and Sheth, A. P., 2018, 'Machine learning for Internet of Things data analysis: a survey', *Digital Communications and Networks*, vol. 4, no. 3, pp. 161–175.

Martínez-Muñoz, G. and Suárez, A., 2010, 'Out-of-bag estimation of the optimal sample size in bagging', *Pattern Recognition*, vol. 43, no. 1, pp. 143–152.

Müller, A. C. and Guido, S., 2016, *Introduction to machine learning with Python: a guide for data scientists*, O'Reilly Media, Inc.

Neelamegam, S. and Ramaraj, E., 2013, 'Classification algorithm in data mining: an overview', *International Journal of P2P Network Trends and Technology (IJPTT)*, vol. 4, no. 8, pp. 369–374.

Nguyen, L., 2017, 'Tutorial on support vector machine', *Applied and Computational Mathematics*, vol. 6, no. 4–1, pp. 1–15.

Pagel, J. F. and Kirshtein, P., 2017, *Machine dreaming and consciousness*, Academic Press, New York, NY.

Pedregosa, F., Varoquaux, G., Gramfort, A., Michel, V., Thirion, B., Grisel, O., Blondel, M., Prettenhofer, P., Weiss, R., Dubourg, V. and Vanderplas, J., 2011, 'Scikit-learn: machine learning in Python', *Journal of Machine Learning Research*, vol. 12, pp. 2825–2830.

Probst, P. and Boulesteix, A. L., 2017, 'To tune or not to tune the number of trees in random forest', *Journal of Machine Learning Research*, vol. 18, pp. 181–181.

Quinlan, J. R., 1986, 'Induction of decision trees', *Machine Learning*, vol. 1, no. 1, pp. 81–106.

Quinlan, J. R. C., 1993, *Programs for machine learning*, vol. 4, Morgan Kaufmann Publishers, p. 5.

Raschka, S., 2018, 'Model evaluation, model selection, and algorithm selection in machine learning', *arXiv preprint arXiv:1811.12808*.

Samuel, Arthur L., 1959, 'Some studies in machine learning using the game of checkers', *IBM Journal of Research and Development*, vol. 44, pp. 206–226.

Shanmuganathan, S., 2016, 'Artificial neural network modelling: an introduction', in *Artificial neural network modelling*, Springer, Cham, pp. 1–14.

Simeone, O., 2018, 'A brief introduction to machine learning for engineers', *Foundations and Trends® in Signal Processing*, vol. 12, no. 3–4, pp. 200–431.

Stamp, M., 2017, *Introduction to machine learning with applications in information security*, Chapman and Hall/CRC Press, Boca Raton, FL.

Sugiyama, M., 2015, *Introduction to statistical machine learning*, Morgan Kaufmann, Burlington, MA.

Tarca, A. L., Carey, V. J., Chen, X. W., Romero, R. and Drăghici, S., 2007, 'Machine learning and its application in biology', *PLOS Computational Biology*, vol. 3, no. 6, p. e116.

Varma, S. and Simon, R., 2006, 'Bias in error estimation when using cross-validation for model selection', *BMC Bioinformatics*, vol. 7, no. 1, p. 91.

Witten, I. H., Frank, E., Hall, M. A. and Pal, C. J., 2016, *Data mining: practical machine learning tools and techniques*, Morgan Kaufmann, Burlington, MA.

Wright, M. N. and Ziegler, A., 2017, 'ranger: a fast implementation of random forests for high dimensional data in C++ and R', *Journal of Statistical Software*, vol. 77, no. 1, pp. 1–17.

Wright, M. N., Dankowski, T. and Ziegler, A., 2017, 'Unbiased split variable selection for random survival forests using maximally selected rank statistics', *Statistics in Medicine*, vol. 36, no. 8, pp. 1272–1284.

2

Dimension Reduction Techniques

Muhammad Kashif Hanif, Shaeela Ayesha and Ramzan Talib

CONTENTS

2.1 Dimension Reduction

High dimensional and multidimensional data is found in different applications for object classification, face image recognition, processing of electrocardiography (ECG) signals, medical images classification of disease, telecommunication, process and program monitoring. The preprocessing of high- and multidimensional data is required before applying analysis and machine learning algorithms. When dealing with multidimensional and high-dimensional data the following questions are generally raised: How can one reduce a data set having hundred or even thousands of features while preserving its core properties for analysis? How to visualise a data set having countless dimensions? How to eliminate redundant and noisy features? How to make precise classifications and accurate predictions in the presence of many redundant and missing features (Amaratunga and Cabrera 2016)?

There is need to reduce dimensions of high-dimensional and multidimensional data. Luckily, a series of strategies such as data compression, aggregation, normalisation and dimension reduction are available to accomplish this task. This chapter discusses a variety of dimension reduction techniques. The core focus of this chapter is to identify key features, strengths and limitations of these techniques.

A common goal of dimension reduction is to prepare data for machine learning algorithms and models to make classification and predictions precise. Another objective is to examine the data in situations where classification is unknown and can occur in the future, with the goal of predicting what will be classified and how it can be classified. The performance of machine learning algorithms can be improved when a large number of objects can be classified using few features. For example, rather than dealing with thousands of features, these features can be categorised into small groups and separate models for each group can be built.

The process of reducing the number of features by eliminating redundant and duplicate features is known as dimension reduction. Dimension reduction is commonly applied at the preprocessing stage before applying machine learning models. Dimension reduction can improve data exploration and visualisation (Meng et al. 2016; Radüntz et al. 2017).

Formally, dimension reduction transforms the high dimensional data $Y \in R^{m*h}$ having h dimensions and m observations into low-dimensional data $Z \in R^{m*l}$ where $l \ll h$ in an ideal case. DRT can have different mappings to reconstruct a sample from the low-dimensional representation (Kraemer et al. 2018).

2.2 Dimension Reduction Techniques

Dimension Reduction Techniques (DRTs) create a new representation of the original data by transforming it to lower dimensions with the aim of preserving the original structure of the data. These techniques help to identify sets of redundant and irrelevant features for subsequent analysis. Use of DRTs can reduce the computation time and storage space requirements. Dimension reduction can be performed using feature selection and feature extraction.

2.2.1 Feature Selection

The objective of feature selection is to select a feature set that can precisely define the input with a selected feature set. Feature selection can be performed either by selection of a minimum subset of features or by ranking the features according to a certain criterion or by selecting the top features. Feature selection provides a variety of models, search strategies, quality measures and evaluation to reduce time and improve performance (Liu and Motoda 2007).

Feature selection methods can be classified into filter, wrapper and embedded methods (Elghazel et al. 2016). The filter method uses feature search and feature selection criteria to select a subset of variables based on the intrinsic property of data. The major challenge is redundancy and defining feature relevance. Wrapper methods incorporate clustering algorithms for subset selection of variables. These methods consist of search component, clustering algorithm and criteria for feature evaluation (Liu and Motoda 2007). Wrapper approaches make computationally intensive and classifier-dependent selection using sequential forward selection, sequential backward elimination and genetic algorithm. The embedded method uses the properties of filter and wrapper methods. Embedded methods are implemented by algorithms that have their own predefined feature selection method (e.g., decision trees).

2.2.2 Feature Extraction

Feature extraction transforms original feature sets into fewer dimensions. There exist a wide variety of DRTs that can perform low-dimension transformations. These transformations can be linear or non-linear.

2.3 Linear Dimension Reduction Techniques

Linear Dimension Reduction Techniques (LDRTs) perform linear transformations. For this purpose, an explicit linear transformation function is used to decrease the dimensionality and increase the efficiency and relevance for the given application (Cunningham and Ghahramani 2015). This section presents the linear DRTs.

2.3.1 Principal Component Analysis

Principal Component Analysis (PCA) is used by a large number of applications for feature extraction and dimension reduction. It computes the reduced dimensions based on different distance measures. PCA transforms the original feature set into a new feature set by using linear transformations.

The new set of derived features is known as Principal Components (PCs). The main objective of the PCA method is to find that the data represented in high dimensions can be effectively modeled with fewer dimensions. Mathematically, PCA works by taking an eigenvalue decomposition of the covariance or a correlation matrix. Here, the eigenvalues can represent the total variance explained by the corresponding eigenvalues of PCA (Jolliffe and Cadima 2016; Meng et al. 2016).

PCA can be performed using the following steps:

- Find the mean vector m.
- Compute the covariance matrix.
- Compute the eigenvector and the corresponding eigenvalues.
- Perform ranking and choosing the top k eigenvectors.
- Transform data samples to a new subspace.

PCA attempts to rotate the first component in the direction of the largest variance. The important question is to decide how many PCs are important to represent data in lower dimensions. There exist different approaches to consider the number of PCs to retain. For example, the Kaser-Guttman method simply states that PCs with a value greater than 1 eigenvalue should be retained.

The dimensions of a reduced feature set must be small enough to effectively reduce the dimensions of data and large enough to retain the maximum amount of information to represent the original data. The quality of these representations can be measured using square distance between the vectors and their projections in reduced feature set. The aim of PCA is to identify the projection matrix that minimises the loss function (Erichson et al. 2018). Real-world applications of PCA include face recognition, image compression, pattern recognition, data integration, data analysis and visualisation.

2.3.2 Singular Value Decomposition

Singular Value Decomposition (SVD) provides a low-dimensional representation of high-dimensional data. SVD was specifically developed for matrix decomposition. It factorises matrices into two orthogonal matrices and a diagonal matrix. SVD follows a *de facto* standard for dimension reduction in generic data sets. SVD can be applied on any real-world data that can be presented as matrix. It provides an optimal level of dimension reduction based on linear projections (Erichson et al. 2016). SVD has been applied in various applications like natural language processing (Mills 2017), bio-informatics and signal processing (Wang and Zhu 2017).

To work with orthogonal and sparsity issue, Guillemot et al. (2019) proposed constrained SVD that can incorporate multiple constraints to classical SVD. This approach can be used for efficient preprocessing and management of data collected from multiple sources having a diverse nature. Husson et al. (2019) introduced a multi-level SVD-based imputation method. This new approach can be useful in different fields such as education, medical and life science where the multi-source and mixed nature of data is found extensively.

2.3.3 Latent Discriminant Analysis

Latent Discriminant Analysis (LDA) is a popular LDRT, which works similar to PCA to reduce the original feature set into lower dimensions. PCA focuses on component axes to reduce variance of data, while LDA aims to minimise the differences between classes. For example, two classes are not separable if one projects the data onto the y-axis (e.g., overlap). In the real-world application, the relationship between the classes will be certainly more complex, and dimensionality will be much higher.

LDA is a supervised learning technique that performs linear transformations. LDA works on axes that reduce the separation between multiple classes. LDA captures the global structure of data and ignores the geometrical variation of local data points of the same class. Real-world applications that apply LDA for dimension reduction include machine learning, computer vision and image classification.

2.3.4 Independent Component Analysis

Independent Component Analysis (ICA) is an unsupervised learning approach for exploration of multi-channel data. In its classical form, ICA models the data as a linear mixture of non-Gaussian independent sources (Ren et al. 2018). ICA is closely related to PCA and Factor Analysis (FA). ICA was mostly used to identify unique components from mixed signals and to eliminate irrelevant features to reduce dimensionality, while preserving the original form of signals (Jayaprakash et al. 2018).

ICA is a multi-variate approach used to perform linear transformation in such a way that the resulting components are uncorrelated and independent. ICA can be used as an analytical and computation method for finding hidden features. In ICA, data features can be formed from the linear mixture of some latent features. The latent variables are followed by non-Gaussian and mutually independent factors, known as Independent Components (ICs) of the data.

Popular applications of ICA include signal processing, continuous data analysis, compression, Bayesian detection, source localisation, blend source separation and identification and mixed source identification and separation (Rahmanishamsi et al. 2018). Radüntz et al. (2017) introduced an IICA-based feature extraction method for automatic EEG artifact elimination. ICA has shown viable performance for signals data processing and identification and separation of irrelevant sources of vibration (Ren et al. 2018).

Application of ICA becomes limited for nonlinear data transformations. It is difficult or impossible to estimate for Gaussian representations using ICA. The benefits of using ICA are:

- Blend source separation in image and signals data.
- No need for labeled data for transformations.
- Removal of noise from the global structure and preservation of the global structure of signals.
- Easier to interpret.
- Explanation of the maximum amount of variance using fewer components.
- Provides a more accurate modeling of non-Gaussian components thus improving estimation.

2.3.5 Projection Pursuits

Projection Pursuits (PP) is a simple and flexible supervised learning dimension reduction approach. It has been extensively used for exploratory data analysis of multivariate data. Originally, PP was a non-parametric technique that found low-dimensional projections for linear data and explored interesting patterns for analysis (Jiang et al. 2015). PP uses a Projection Index (PI) that does not require prior knowledge, which can make PP fast. PP aims to preserve the ordering of features in lower dimensions.

PI has been used to measure the level of interestingness in patterns. Moreover, PI makes PP flexible for various pattern recognition tasks such as cluster analysis and classification (Espezua et al. 2015). Projections in each group remain independent of each other. Parametric versions of PP can be used to maximise the value of PI based on parallel parametric PP and sequential parametric PP. Real-world applications of PP include hyperspectral image analysis, image and non-image data classification, multivariate and exploratory data analysis.

2.3.6 Latent Semantic Analysis

Latent Semantic Analysis (LSA) is an unsupervised linear mapping designed for text documents. It is based on the PCA or SVD computation. It is used to eliminate redundant features and preserve the semantic structure of documents in reduced representation. LSA was developed for information retrieval, especially when, from a huge collection of documents, only few relevant documents match with a given query. LSA is a vector-based technique that has been used to compare and represent the text of a high-dimensional corpus into lower dimensions (Dokun and Celebi 2015).

With an explosive growth of text documents, term similarities in different fields, and their associations, are creating many issues. For example, in a report there is a need to collect document forms (i.e., health, sports, Olympic games, and viruses etc.). In such a situation, a term may cause conflict in identification and classification (i.e., "fly" and "fly" both have the same spellings but one can be used as a noun and the other as a verb) (Talib et al. 2016). To overcome such issues and improve the analysis of text data, Elghazel et al. (2016) introduced an ensemble-based multi-label classification method.

Other applications of LSA include searching words and their meanings, finding pairs of synonyms and antonyms, educational applications such as assessment of essays, to analyse the competency pattern of professionals for successful business management (Müller et al. 2016) and the communication between patient and physician for medication (Vrana et al. 2018). LSA can also be applied to images and videos (Gao et al. 2017) and for identification of medical records (Gefen et al. 2018).

2.3.7 Locality Preserving Projection

Locality Preserving Projection (LPP) is an unsupervised LDRT that relies on the linear approximation of nonlinear Laplacian Eigen maps. These graphs can be developed using Laplacian graph notation. Linear projective maps were used to solve variation issues and optimal preservation of neighborhood structure of data. LPP preserves local structure, intrinsic geometry of the data and the local distance between samples when projecting data to lower dimensions. Classical LPP is unable to preserve global and manifold projections. Variants of LLP can preserve global and manifold structures of data during the transformation process. Real-world applications of LPP include image recognition and classification (He 2004).

2.4 Nonlinear Dimension Reduction Techniques

To overcome the limitations of LDRTs, numerous efforts have been devoted to develop nonlinear feature transformations (Chen et al. 2018). This section presents a brief overview of nonlinear techniques.

2.4.1 Kernel Principal Component Analysis

Kernel Principal Component Analysis (KPCA) is an extension of classical PCA to perform nonlinear transformations. Instead of calculating the covariance of a matrix, KPCA calculates the principal Eigen vectors of the kernel matrix. Kernel property makes PCA suitable for nonlinear mapping (Van Der Maaten et al. 2009; Xie et al. 2016). KPCA uses polynomial and Gaussian kernel that makes its functionality limited for manifold learning. Variants of KPCA such as Subset KPCA (SKPCA) were introduced to reduce the computational complexities of KPCA for dimension reduction and classification. Multi-Scale KPCA (MSKPCA) was developed as a fault diagnostic method for nonlinear process monitoring (Chen et al. 2018). Other applications of KPCA include image classification, sensor data, medical and bio-informatics.

2.4.2 Isomap

Isomap is a nonlinear DRT based on unsupervised learning. Isomap aims to preserve pairwise geodesic distances between data points (Najafi et al. 2015). Isomap defines the geodesic distance as a sum of edge weights along the shortest path between two nodes.

Classical scaling computes the pairwise Euclidean distances and does not consider the distribution of the neighboring data points. Classical scaling might consider two data points as near points when the high-dimensional data lies on or near a curved manifold. The distance between these two points over the manifold can be much larger than the interpoint distance. Isomap solves this problem by preserving pairwise geodesic distances between data points (Gao et al. 2017).

Isomap maps the original data to the reduced feature set while preserving manifold characteristics. The top n eigenvectors of the geodesic distance matrix represent the coordinates in the new n-dimensional Euclidean space. A common Isomap algorithm consists of the following steps:

- Construct a neighborhood graph G for each point y_i.
- Calculate the shortest path using Dijkstra's or Folyd's algorithms.
- Find the low-dimensional representation of the input data using multidimensional scaling.

Feature management becomes computationally expensive for a large number of feature sets. Isomap can be applied to detect irregularities from large-scale real-time video analytics (Rao et al. 2016), speech summarisation (Liu et al. 2017), to predict the good and bad condition of urban road traffic (Liu et al. 2018), to manage nonlinear data for the identification of cracks in material (Mousavi Nezhad et al. 2018) and to obtain higher compression of video (Li 2018). Huang et al. (2019) proposed Semi-Supervised Discriminant Isomap (SSD-Isomap) to improve visualisation and achieved more precise results for image classification.

2.4.3 Locally Linear Embedding

Locally Linear Embedding (LLE) is an unsupervised nonlinear dimension reduction approach. LLE is considered as a nearest-neighbor approach because it optimally reduces the high-dimensional data by reconstructing data points from a weighted combination of neighbors. LLE, like Isomap, develops graphs to represent data points. LLE works by:

- Assigning a neighbor to each data point.
- Computing the weight that reconstructs the original feature set.
- Low-dimension embedding preserving the geometric locality property.
- LLE finding subspace that best preserves the local linear structure.

LLE does not use class membership for projecting data points in lower dimensions. Hettiarachchi and Peters (2015) proposed a Multiple Manifold LLE (MM-LLE) approach to learning multiple manifolds for multiple classes. MM-LLE has shown better performance and classification accuracy than classical LEE and Isomap. Real-world applications of LLE include image compression, signal processing and image classification.

2.4.4 Self Organising Map

Self Organising Map (SOM) is an excellent technique for visualisation and exploratory data analysis. SOM has been used to find patterns and clusters in the data. SOMs are versatile for analysing nonlinear projections, multi-variate and complex data sets (Kohonen 2013).

Applications of SOM include automatic organisation of a massive collection of documents, intrusion detection etc. Many researchers applied SOMs to perform multidimensional data analysis. For example, SOM was applied to a corporate database to develop risk-based prioritisation (Mounce et al. 2015), recognition of protein folds (Polat and Dokur 2016), for market data analysis (Das et al. 2016), for the classification of fMRI (Bahrami and Shamsi 2017), to remove noise from 6D synthetic spectral images (Merényi and Taylor 2017).

2.4.5 Learning Vector Quantisation

Learning Vector Quantisation (LVQ) is a supervised nonlinear DRT that is similar to SOM. Classification can be performed using distance between input vectors. The class borders enhance the classification accuracy of LVQ. LVQ is suitable for many real-time applications (Yousefi and Hamilton-Wright 2014). LVQ is a competitive-based neural network introduced by Kohonen (2013). LVQ consists of two layers, known as the input and the output layer. LVQ has gained popularity in different application areas due to its classification accuracy.

2.4.6 t-Stochastic Neighbor Embedding

t-Stochastic Neighbor Embedding (t-SNE) is an unsupervised nonlinear DRT used to reduce HDD in low-dimensional space. t-SNE is based on matching distances between distributions. t-SNE is non-parametric and well suited for visualisation of data sets having nonlinear structures (Konstorum et al. 2018). It captures most of the local structure of high-dimension data revealing a global structure and has functionality to work with manifold learning (McInnes et al. 2018). It performs well for image processing, audio and video applications.

2.5 Conclusion and Future Directions

This chapter described a general overview of different dimensional reduction techniques based on linear and nonlinear transformations. The usefulness of a technique can depend on various factors, such as: the size of the data set; the nature of the data (i.e., linear, nonlinear); the types of patterns that exist in the data; the levels of redundancy and noise that exist in the data; if any of the data meet some underlying assumptions of the particular technique; and the particular goal of analysis. Different methods can lead toward different results and their performance may vary with changes in the data set. Therefore, it is necessary to analyse several techniques and select the one that gives optimal results for given parameters to achieve a specific goal. Combination of classical dimension reduction techniques and deep learning can be explored for different applications to improve performance.

References

Amaratunga, D. and Cabrera, J., 2016, 'High-dimensional data', *Journal of the National Science Foundation of Sri Lanka*, vol. 44, no. 1, pp. 3–9.

Bahrami, S. and Shamsi, M., 2017, 'A non-parametric approach for the activation detection of block design fMRI simulated data using self-organizing maps and support vector machine', *Journal of Medical Signals and Sensors*, vol. 7, no. 3, pp. 153.

Chen, J., Wang, G. and Giannakis, G. B., 2018, 'Nonlinear dimensionality reduction for discriminative analytics of multiple datasets', *IEEE Transactions on Signal Processing*, vol. 67, no. 3, pp. 740–752.

Cunningham, J. P. and Ghahramani, Z., 2015, 'Linear dimensionality reduction: survey, insights, and generalizations', *The Journal of Machine Learning Research*, vol. 16, no. 1, pp. 2859–2900.

Das, G., Chattopadhyay, M. and Gupta, S., 2016, 'A comparison of self-organising maps and principal components analysis', *International Journal of Market Research*, vol. 58, no. 6, pp. 815–834.

Dokun, O. and Celebi, E., 2015, 'Single-document summarization using latent semantic analysis', *International Journal of Scientific Research in Information Systems and Engineering (IJSRISE)*, vol. 1, no. 2, pp. 57–64.

Elghazel, H., Aussem, A., Gharroudi, O. and Saadaoui, W., 2016, 'Ensemble multi-label text categorization based on rotation forest and latent semantic indexing', *Expert Systems with Applications*, vol. 57, pp. 1–11.

Erichson, N. B., Voronin, S., Brunton, S. L. and Kutz, J. N., 2016, 'Randomized matrix decompositions using R', *arXiv preprint arXiv:1608.02148*.

Erichson, N. B., Zheng, P., Manohar, K., Brunton, S. L., Kutz, J. N. and Aravkin, A. Y., 2018, 'Sparse principal component analysis via variable projection', *arXiv preprint arXiv:1804.00341*.

Espezua, S., Villanueva, E., Maciel, C. D. and Carvalho, A., 2015, 'A projection pursuit framework for supervised dimension reduction of high dimensional small sample datasets', *Neurocomputing*, vol. 149, pp. 767–776.

Gao, L., Song, J., Liu, X., Shao, J., Liu, J. and Shao, J., 2017, 'Learning in high-dimensional multimedia data: the state of the art', *Multimedia Systems*, vol. 23, no. 3, pp. 303–313.

Gefen, D., Miller, J., Armstrong, J. K., Cornelius, F. H., Robertson, N., Smith-McLallen, A. and Taylor, J. A., 2018, 'Identifying patterns in medical records through latent semantic analysis', *Communications of the ACM*, vol. 61, no. 6, pp. 72–77.

Guillemot, V., Beaton, D., Gloaguen, A., Löfstedt, T., Levine, B., Raymond, N., Tenenhaus, A. and Abdi, H., 2019, 'A constrained singular value decomposition method that integrates sparsity and orthogonality', *PloS one*, vol. 14, no. 3, p. e0211463.

He, X., 2004, 'Incremental semi-supervised subspace learning for image retrieval', in *Proceedings of the 12th annual ACM international conference on multimedia*, ACM, pp. 2–8.

Hettiarachchi, R. and Peters, J. F., 2015, 'Multi-manifold LLE learning in pattern recognition', *Pattern Recognition*, vol. 48, no. 9, pp. 2947–2960.

Huang, R., Zhang, G. and Chen, J., 2019, 'Semi-supervised discriminant Isomap with application to visualization, image retrieval and classification', *International Journal of Machine Learning and Cybernetics*, vol. 10, no. 6, pp. 1269–1278.

Husson, F., Josse, J., Narasimhan, B. and Robin, G., 2019, 'Imputation of mixed data with multilevel singular value decomposition', *Journal of Computational and Graphical Statistics*, vol. 28, no. 3, pp. 1–26.

Jayaprakash, C., Damodaran, B. B. and Soman, K. P., 2018, 'Randomized ICA and LDA dimensionality reduction methods for hyperspectral image classification', *arXiv preprint arXiv:1804.07347*.

Jiang, T., Jia, H., Yuan, H., Zhou, N. and Li, F., 2015, 'Projection pursuit: a general methodology of wide-area coherency detection in bulk power grid', *IEEE Transactions on Power Systems*, vol. 31, no. 4, pp. 2776–2786.

Jolliffe, I. T. and Cadima, J., 2016, 'Principal component analysis: a review and recent developments', *Philosophical Transactions of the Royal Society A: Mathematical, Physical and Engineering Sciences*, vol. 374, no. 2065, p. 20150202.

Kraemer, G., Reichstein, M. and Mahecha, M. D., 2018, 'dimRed and coRanking—unifying dimensionality reduction in R', *R Journal*, vol. 10, no. 1, pp. 342–358.

Kohonen, T., 2013, 'Essentials of the self-organizing map', *Neural Networks*, vol. 37, pp. 52–65.

Konstorum, A., Vidal, E., Jekel, N. and Laubenbacher, R., 2018, 'Comparative analysis of linear and nonlinear dimension reduction techniques on mass cytometry data', *bioRxiv*, p. 273862.

Liu, H. and Motoda, H. (eds.), 2007, *Computational methods of feature selection*, CRC Press, Boca Raton, FL.

Liu, S. H., Chen, K. Y., Chen, B., Wang, H. M. and Hsu, W. L., 2017, March, 'Leveraging manifold learning for extractive broadcast news summarization', in *IEEE international conference on acoustics, speech and signal processing* (ICASSP), pp. 5805–5809.

Liu, Y., Xia, C., Fan, Z., Wu, R., Chen, X. and Liu, Z., 2018, 'Implementation of fractal dimension and self-organizing map to detect toxic effects of toluene on movement tracks of Daphnia magna', *Journal of Toxicology*.

Li, H., 2018, '1D representation of Isomap for united video coding', *Multimedia Systems*, vol. 24, no. 3, pp. 297–312.

McInnes, L., Healy, J. and Melville, J., 2018, 'Umap: uniform manifold approximation and projection for dimension reduction', *arXiv preprint arXiv:1802.03426*.

Meng, C., Zeleznik, O. A., Thallinger, G. G., Kuster, B., Gholami, A. M. and Culhane, A. C., 2016, 'Dimension reduction techniques for the integrative analysis of multi-omics data', *Briefings in Bioinformatics*, vol. 17, no. 4, pp. 628–641.

Merényi, E. and Taylor, J., 2017, 'Som-empowered graph segmentation for fast automatic clustering of large and complex data', in *2017 12th international workshop on self-organizing maps and learning vector quantization, clustering and data visualization (WSOM)*, IEEE, pp. 1–9.

Mills, P., 2017, October 5, 'Singular value decomposition (svd) tutorial: applications, examples, exercises a complete tutorial on the singular value decomposition method', https://blog.statsbot.co/singular-value-decomposition-tutorial-52 c695315254.

Mounce, S. R., Sharpe, R., Speight, V., Holden, B. and Boxall, J., 2015, 'Self-organizing maps for knowledge discovery from corporate databases to develop risk based prioritization for stagnation', in *11th International Conference on Hydroinformatics*.

Mousavi Nezhad, M., Gironacci, E., Rezania, M. and Khalili, N., 2018, 'Stochastic modelling of crack propagation in materials with random properties using isometric mapping for dimensionality reduction of nonlinear data sets', *International Journal for Numerical Methods in Engineering*, vol. 113, no. 4, pp. 656–680.

Müller, O., Schmiedel, T., Gorbacheva, E. and Vom Brocke, J., 2016, 'Towards a typology of business process management professionals: identifying patterns of competences through latent semantic analysis', *Enterprise Information Systems*, vol. 10, no. 1, pp. 50–80.

Najafi, A., Joudaki, A. and Fatemizadeh, E., 2015, 'Nonlinear dimensionality reduction via path-based isometric mapping', *IEEE Transactions on Pattern analysis and Machine Intelligence*, vol. 38, no. 7, pp. 1452–1464.

Polat, O. and Dokur, Z., 2016, 'Protein fold recognition using self-organizing map neural network', *Current Bioinformatics*, vol. 11, no. 4, pp. 451–458.

Radüntz, T., Scouten, J., Hochmuth, O. and Meffert, B., 2017, 'Automated EEG artifact elimination by applying machine learning algorithms to ICA-based features', *Journal of Neural Engineering*, vol. 14, no. 4, p. 046004.

Rahmanishamsi, J., Dolati, A. and Aghabozorgi, M. R., 2018, 'A copula based ICA algorithm and its application to time series clustering', *Journal of Classification*, vol. 35, no. 2, pp. 230–249.

Rao, A. S., Gubbi, J. and Palaniswami, M., 2016, 'Anomalous crowd event analysis using isometric mapping', in *Advances in signal processing and intelligent recognition systems*, Springer, Cham, pp. 407–418.

Ren, W., Wen, G., Luan, R., Yang, Z. and Zhang, Z., 2018, 'Single-channel blind source separation and its application on arc sound signal processing', in *Transactions on intelligent welding manufacturing*, Springer, Singapore, pp. 115–126.

Talib, R., Hanif, M. K., Ayesha, S. and Fatima, F., 2016, 'Text mining: techniques, applications and issues', *International Journal of Advanced Computer Science and Applications*, vol. 7, no. 11, pp. 414–418.

Van Der Maaten, L., Postma, E. and Van den Herik, J., 2009, 'Dimensionality reduction: a comparative', *Journal of Machine Learning Research*, vol. 10, no. 66–71, p. 13.

Vrana, S. R., Vrana, D. T., Penner, L. A., Eggly, S., Slatcher, R. B. and Hagiwara, N., 2018, 'Latent semantic analysis: a new measure of patient-physician communication', *Social Science and Medicine*, vol. 198, pp. 22–26.

Wang, Y. and Zhu, L., 2017, May, 'Research and implementation of SVD in machine learning', in *2017 IEEE/ACIS 16th international conference on computer and information science (ICIS)*, pp. 471–475.

Xie, L., Li, Z., Zeng, J. and Kruger, U., 2016, 'Block adaptive kernel principal component analysis for nonlinear process monitoring', *AIChE Journal*, vol. 62, no. 12, pp. 4334–4345.

Yousefi, J. and Hamilton-Wright, A., 2014, 'Characterizing EMG data using machine-learning tools', *Computers in Biology and Medicine*, vol. 51, pp. 1–13.

3

Reviews Analysis of Apple Store Applications Using Supervised Machine Learning

Sarah Al Dakhil and Sahar Bayoumi

CONTENTS

3.1 Introduction

Application distribution platforms such as Apple App Store and Google Play enable users to search, buy and install software applications with a few clicks (Pagano and Maalej, 2015). Nowadays, people from various age groups are utilising the app store on a daily basis. Additionally, application developers are working in different directions, which leads to the production of a variety of applications. This wealth of applications enables us to deliver analytical studies to serve different fields and provide valuable feedback to developers. Regardless of whether researchers have any access to the applications' content or not, there are attributes or features that can serve the researchers' needs (Aiquwayfili et al., 2013). And now, as we see, Apple's iTunes App Store, as one of the first contenders, has become and continues to be the leader of the app store market. Apple has grown rapidly to exceed 200,000 applications per month (Kimbler et al., 2010).

Studies mention the importance of reviews for the success of an app (Pagano et al., 2013). Apps with better reviews get a better rating in the store and with it better visibility and higher sales and download numbers (Maleej et al., 2015). The reviews seem to help users navigate apps and decide which one to use. Moreover, recent research has pointed to the potential importance of reviews for app developers as well as vendors. As demonstrated in Agrawal et al. (2018), evaluation of numerous reviews posted by users on e-commerce and social networking is of considerable use, as it contains highly rated information. Therefore, sentiment analysis and opinion mining are active research of the NLP domain as well as in artificial intelligence.

The success of the company or product directly depends on its customers. If the customer is impressed with your product, then the product is a success. If not, then you need to make some changes. So how you will know if your product is a success or not? You need to analyze your customers, and this is where sentiment analysis come into the picture. Sentiment analysis is the process of computationally identifying and categorising opinion from customer reviews and it determines the customers' attitude toward the product whether positive, neutral or negative.

In the real world, an organisation needs to take a good decision every other minute, in order to ensure organisational success, decisions such as introducing a new product and exploring potential demand for it, its market share and profitability, and the competitor's benchmark in the market and so on (Kalra et al., 2019). This research project's major contribution is helping developers in their evolution of apps by extracting the hidden patterns and features that are exhibited in user reviews from customers within the iTunes website as well as helping app developers make successful decisions.

The chapter will discuss:

(1) Providing techniques for classifying the reviews based on their metadata (star rating), keyword frequencies and sentiment analysis.
(2) Comparing the accuracy of multiple Supervised ML classification algorithms.
(3) Utilising feature extraction algorithms combined with Supervised Machine Learning algorithm and natural language processing (NLP) to add strength in the classification of, and then prediction for, text reviews.

The remainder of the chapter is structured as follows. Related work is reviewed in Section 2. Section 3 describes the project implementation and design, including data collection and methodology, that was adapted for this study. Section 4 reports on the results, comparing the accuracy and the performance of both feature extraction (FE) algorithms combined with the variety of machine learning (ML) classification algorithms, and the findings are discussed. Future work and recommendations are discussed in Section 5. Finally, Section 6 concludes the chapter.

3.2 Literature Review

Customer satisfaction has been the fundamental pointer for companies to improve service, particularly for long-term profit, customer loyalty and customer retention (Yang et al., 2017). The major reason for measuring customer satisfaction is to get the information from customers for the end goal of administrative decision-making to maximise and keep customers satisfied (Yang et al., 2017).

Therefore, the related studies focused on machine learning algorithms and feature extraction algorithms in the online reviews, whether for products or for services that focused only on the application store for smart phones.

3.2.1 Machine Learning Algorithms

User reviews on mobile application stores represent a rich source of technical information for app developers (Guzman, 2014). Agrawal and Batra (2013) illustrated the importance of text mining in machine learning which is an emerging technology for web data mining. Then, the authors mention that the goal of text mining strategies is to reduce the effort required of users to obtain useful information from large computerised text data sources.

Guzman et al. (2015) proposed an ensemble of machine learning classifiers in order to classify user reviews into seven categories relevant for software evolution. These categories are: bug reports, feature strength, feature shortcoming, user request, praise, complaint, usage scenario. They tested the machine learning algorithms (i.e., Naïve Bayes, Support Vector Machines (SVMs), Logistic Regression and Neural Networks) on 4,550 reviews mined from the Apple App Store. Then they mentioned the authors (Guzman et al., 2015); they believe that their approach can be extended to automatically rank the categorised reviews by taking ratings and sentiments contained in the review comments into consideration. Moreover, the mechanisms to summarise and visualise the content of the classified reviews could further reduce the processing effort.

According to Ali et al. (2017), the goal of their study is to semi-automatically classify the user reviews of the app-pairs and compare them at the app and platform level. To achieve this, the authors use both natural language processing and semi-supervised machine learning algorithms, which were Naïve Bayes (NB) and Support Vector Machines (SVM) implemented by use of the Scikit Learn Tool, to evaluate the accuracy of the classifiers; they measured the F-measure for both of the Naïve Bayes and SVM algorithms, and they found that SVM achieves a higher F-measure than Naïve Bayes. One of the main results from this study (Ali et al., 2017) demonstrates that on average, iOS apps receive more critical and post update complaints while Android apps receive more complaints related to app features and non-functional properties.

Another study in 2018 conducted by Jha et al. (2018) explored the performance of a novel semantically aware methodology for classifying and summarising user reviews on app stores. The proposed methodology depends on semantic role labelling. So, individual user review sentences are extracted and annotated to identify the semantic roles played by the words that appear in each sentence. They use Support Vector Machines (SVM) and Naïve Bayes (NB) by WEKA tool and have reported their superiority over other classifiers in short-text classification tasks. Before the end of this study (Jha et al., 2018), recall, precision and the F-score are used to evaluate the performance of the different classification techniques used in their analysis.

In another study targeting only one-star and two-star reviews, McIlroy et al. (2016) studied the extent of multi-labeled user reviews (reviews raising more than one issue type) and proposed an approach to automatically labelling multi-labeled user reviews in the Google Play Store and Apple App Store. They defined 13 types of issue and labeled a number of reviews manually to form their gold standard dataset. For labeling tasks, they transformed the problem of multi labeling into single labeling and used a classifier for each label and combined their results. They used several different classifiers e.g., support vector machines (SVM), decision tree (J48) and Naïve Bayes (NB) as well as several different multi-labeling approaches, e.g., Binary Relevance (BR) (it does not leverage the correlations between labels), and Pruned Sets with threshold extensions (PSt). They defined a threshold to assign each label to a review. Finally, they used ten-fold cross-validation to evaluate results. In the pre-processing phase, they removed numbers and special characters, but not stop words, expanded abbreviations, filtered words occurring less than three times in the dataset, stemmed words, and removed reviews consisting of three words or less. However, observations exhibit that reviews with less than three words report bugs and issues as well (e.g., "poor camera," "save button sucks," and "can't upload picture"). Moreover, they used TfidfVectorizer (TF-IDF) as a mean to increase the weight of words that occur frequently in a single user review and to decrease the weight of words that occur frequently in many user reviews. Although it helps to devalue ordinary words, this way of weighting words might demote issues repeated and discussed between several users.

3.2.2 Feature Extraction Algorithms

Reviews written by the users in app stores are considered a valuable source of information for app vendors and developers, as they include information about bugs, ideas for new features, or documentation of released features. In 2015, Walid et al. proposed a probabilistic technique to classify app reviews into four categories: bug reports, feature requests, user experiences and ratings. First, they collected real reviews from app stores and extracted their metadata. They use review metadata such as the star rating and the tense, as well as text classification, natural language processing and sentiment analysis techniques. Then they conducted a series of experiments to compare the accuracy of the techniques and compared them with simple string matching. They generated a list of keywords by string matching, bag of words, sentiment scores, NLP pre-processed text, review rating and length, to be used for classification task. Then, they applied Naïve Bayes, Decision Trees and MaxEnt (maximum entropy) to compare the performance of binary to multiclass classifiers in classification of user feedback into the predefined basic types.

The authors Maleej et al. (2016) extended their approach by adding bigram and its combinations to utilised classification techniques, and by improving pre-processing phases and classification scripts. They argued that by the use of metadata combined with text classification and natural language pre-processing of the text, the classification precision rises significantly. They found that metadata alone results in poor classification accuracy. When combined with natural language processing, the classification precision got between 70% and 95% while the recall got between 80% and 90%. Therefore, text classification should be enhanced with metadata such as the tense of the text, the star rating, the sentiment score and the length. The results show that app reviews can be classified as to bug reports, feature requests, user experiences and ratings (praise or dispraise) with a high accuracy of between 70% and 97%. Complementary within-app analytics such as feature extraction, opinion mining and summarisation of the reviews, will make app store data more useful for decisions about software and engineering requirements.

Harman et al. presented app store mining and analyzed technical and business aspects of apps by extracting app features from the official app descriptions. Panichella et al. (2015) presented a taxonomy to classify app reviews into categories in order to support software maintenance and evolution. Also, they designed an approach that integrates text analysis, sentiment analysis and natural language processing techniques to automatically classify app reviews into the proposed categories. Their approach consists of four steps: first, designing a taxonomy for software maintenance and evolution that can lead developers to select the reviews most useful for a specific maintenance task (i.e. bug fixing, feature adding, etc.); second, extracting features from user reviews data by using the tf (term frequency) as a feature extraction algorithm, NLP, text analysis and sentiment analysis techniques;

third, learning classifiers by using the features extracted to train ML algorithms, which were Naïve Bayes classifier, Logistic Regression, Support Vector Machines, J48 and the Alternating Decision Tree (ADTree); finally, evaluating the performance of the ML techniques experimented in the previous step. The results of the evaluation showed that a combination of the three techniques had better performance than each technique used individually.

Based on the literature review, we found that very few previous studies have worked on sentiment analysis in the mobile application store online reviews and combined it as a feature with other features and applied classification algorithms in machine learning.

Based on the literature review, we found that Support Vector Machines, Logistic Regression and Naïve Bayes gives high accuracy in the previous work of data mining . Therefore, we will use all these algorithms in addition to the three other algorithms (Random Forest, AdaBoost and Neural Network) that are proposed in this study and compare the results at the end of this project.

3.3 Proposed Methodology

In the full research methodology, the Python programming language was used with a number of libraries. Table 3.1 summarises the most important libraries or packages that were used for this project.

TABLE 3.1

Summarize the Main Libraries or Packages Used in the Project

Software	☐ Python (3.7) version, with Jupyter Notebook environment.
	☐ For running the Jupyter Notebook, this need to install Anaconda which is (The Enterprise Data Science Platform).
Main	☐ Pandas
Packages	☐ Numpy
/	☐ Re
Library	☐ Wordcloud
	☐ NLTK
	☐ PorterStemmer
	☐ WordNetLemmatizer
	☐ Matplotlib
	☐ Statistics
	☐ wordcloud (WordCloud)
	☐ Skleran
	☐ Sklearn.feature_extraction.text (TfidfVectorizer,CountVectorizer)
	☐ Sklearn.model_selection (train_test_split)
	☐ sklearn.metrics (classification_report)
	☐ Machine Learning libraries from Skleran

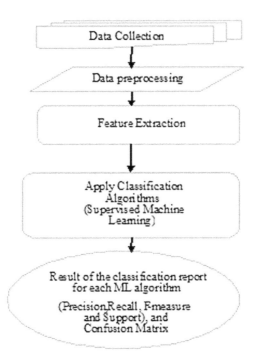

FIGURE 3.1
Methodology steps.

The methodology of this project goes through four stages: data collection, data pre-processing, feature extraction and Supervised Machine Learning Classification. Figure 3.1 shows and summarise these steps.

3.3.1 Data Collection

The data was extracted from the iTunes website (https://itunes.apple.com/) at the Apple Inc website. Multiple of Virtsual Linux Server (VPS) was used within Python script by importing the open source library BeautifulSoup* that is used for web scraping. BeautifulSoup is a Python HTML parsing library, which can be combined with language native support for HTTP connections, whereas web scraping usually refers to the automated process of data transfer between a web page and a structured data format implemented using a toolkit called a web robot, scraper or crawler (Gheorghe, 2018). VPS stands for Virtual Private Server, the paid servers (5 VPN† servers from the type Rise-1) that were used while running on the Linux system to have high performance. In addition, it also needs the high-speed internet up to 500 Mbps.

---

* BeautifulSoup – www.crummy.com/software/BeautifulSoup/
† VPN "Rise-1" – www.ovh.com/world/dedicated-servers/rise/rise-1/

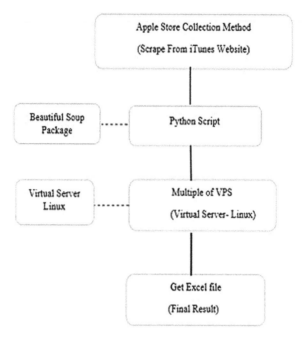

FIGURE 3.2
Dataset collection method.

That is why these servers were hired. The main steps for data collection are shown in Figure 3.2.

The dataset includes 12 general attributes, which are: Apple URLs, Application Name, Brief Description, Overall Ratings, Number of Ratings, Category, Price, Age Rating, Languages, Current Version Number, What's New and Application Website. In addition to the three user group attributes, each of them includes five features (username, date, rate, title and review text). The dataset focuses on six categories (Business, Education, Games, Lifestyle, Social Networking and Utilities) where 4,000 applications were collected for each category, bringing the total for the dataset to 24,000 records. The sample of our dataset is GitHub,* with an attempt to get three comments from three different users for each application in all datasets, taking into account that each application is unique and not duplicated in all datasets.

3.3.2 Feature Extraction

Feature Extraction (FE) in general is any algorithm that transforms raw data into features that can be used as input for a learning algorithm. In this research chapter the importance of feature extraction combined with ML is due to fact that no ML Algorithm can produce a stable result without feature

---

* Sample of dataset – https://github.com/SarahAlDakhil/Apple-Store-Dataset-Sample-

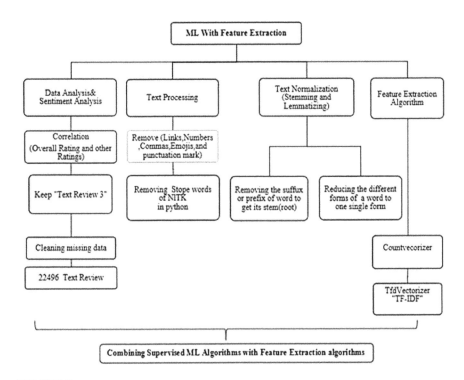

FIGURE 3.3
The main steps of (Machine Learning with Feature Extraction) approach.

engineering. If features are extracted well, then in some cases linear methods show great results. Based on that, in this methodology pre-processing steps were done before feature extraction algorithms and ML classification models, as shown in Figure 3.3, which reflects the main steps of our approach.

The four main steps before applying Supervised ML are:

(A) Data analysis and sentiment analysis.
(B) Text processing.
(C) Text normalisation.
(D) Feature extraction algorithm.

Then, Supervised ML Classification is applied to predict text reviews as positive, negative or neutral). This will be discussed in the following subsection:

3.3.3 Data Analysis and Sentiment Analysis

First, we take Rating 3, and Text Review 3 – which was last reviewed by third user after computing the correlation between Overall Rating and (Text Review 1, Text Review 2 and Text Review 3) – for all categories and

TABLE 3.2

Summarize the Main Libraries or Packages Used in the Project

	Overall Rating	Rating1	Rating2	Rating3
Overall Rating	1	0.420722	0.4236	0.452332
Rating1	0.420722	1	0.326871	0.301638
Rating2	0.4236	0.32687	1	0.36157
Rating3	0.452332	0.301638	0.35617	1

exclude any other attribute, by removing links, emojis, numbers, punctuation marks, commas and stop words in the Text Review 3. Stop words are common English words such as "the," "am," "their" which do not influence the semantics of the review. Removing them can reduce noise and improve the accuracy of machine learning classifiers. While in the second and third steps, we apply Text processing and Text Normalisation using NLTK Python library. This will be discussed in detail in the second and third subsection.

The cross-correlation coefficient was created in order to quantify the level of cross-correlation between unstationary time series (Zebende et al., 2013). The cross-correlation coefficient was calculated between "Overall Rating" with "Rating 1," "Rating 2" and "Rating 3." The results are close to each other, but "Rating 3" shows as more correlated than other ratings with "Overall Rating". Based on that, as mentioned previously, this project will focus on Rating 3 with Text Review 3 – latest review – from the private dataset that is related to this research project as shown in Table 3.2.

Then we focus on Text Review 3. Because not all applications include third review/comment by a third user, we removed the missing data until the dataset record contained 22,496 records for Rating 3, Text Review 3. After preparing the text review by cleaning it, we did a sentiment analysis with compensation for the column "Rating 3", which contains: Negative for values [1,2], Positive for values [4,5] and Neutral for value 3. The total number of negatives in the dataset after we applied the sentiment analysis equals 14,104; positives equal 6,245; while neutrals equal 2,147, as shown in Table 3.3.

TABLE 3.3

Sentiment Analysis with Compensation for Column "Rating 3"

#	Text Review 3	Sentiment
0	Slideshow is too fast and unnecessary. Users c ...	negative
1	I never give reviews but after my experience ...	positive
2	What a useful app \n It's fit to anyone chi ...	positive
3	Like the idea of this app but this to impr ...	neutral
4	Thank Goodness someone thought of a mobile cel...	positive

Text Processing

First, we take Rating 3, and Text Review 3 – last review by third user, for all categories and exclude any other attribute, by removing links, emojis, numbers, punctuation marks, commas and stop words. Stop words are common English words such as "the," "am," "their", which do not influence the semantics of the review. Removing them can reduce noise and improve the accuracy of machine learning classifiers. We have applied Text processing and Text Normalisation for (Stop word removal, stemmer, and lemmatiser) using Natural Language Processing Tool (NLTK). NLTK is used for word tokenisation; POS (Part of Speech) – Tagging, Lemmatisation and Stemming. NLTK is a natural language toolkit designed for symbolic and statistical processing of datasets and language developed by the NLTK team. NLTK was created in 2001 as a part of the Computational Linguistic Department at the University of Pennsylvania.

Before we can use the texts to classify them, we need to clean them a little before classifying them and use both machine learning and feature extraction, so we will remove anything that is not helpful for classification. For example, numbers are not useful to know the sentiment, nor are links, nor special expressions like ";!? /] ^ ... etc., so we will remove all of these things to have clean texts that will be used in the classification, and change everything to lowercase as well by applying the function "clean" that defines for every row of our dataset. To clean all the texts we use the "apply" function from "pandas" package. The detail is given in Table 3.4.

After processing the text by cleaning the text review 3 by removing any links and URLs, stop words, any numbers, we also remove special characters or regular expression likes ; /: | ... etc. and apply sentiment analysis classification, as mentioned earlier. We applied a WordCloud to represent and visualise the most important classification of sentiment analysis, which was done previously. WordClouds provide a simple and effective means to visually

TABLE 3.4

The New Dataset after Cleaning

#	Text	Sentiment	Clean Text
0	Slideshow is too fast and unnecessary, Users c…	Negative	Slideshow fast unnecessary users swipe pace a..
1	I never give reviews but after my experience i…	Positive	Never give reviews experience looking app capt…
2	What a useful app \n It's ft. to anyone chi…..	Positive	Useful app fit anyone child adult app really o…
3	Like the idea of this app but this to impr……..	Neutral	Like idea app improves please least adds inks e…
4	Thank Goodness someone thought of a mobile cel...	Positive	Thank goodness someone thought mobile cell pho..

Note: The column "clean_text" contains the text after removing unnecessary elements

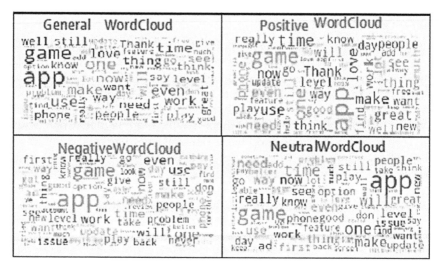

FIGURE 3.4
WordCloud classification.

communicate the most frequent words of text documents, however, they can serve as starting points for deeper text analyses (Lohmann et al., 2015).

Here is the demonstration of WordCloud for the four categories (general "from whole text review in dataset," positive "just with positive sentiment analysis classification," negative "just with negative sentiment analysis classification" and neutral "just with neutral sentiment analysis classification") in the whole dataset that concentrated on just text review 3 for this project as shown in Figure 3.4.

3.3.4 Text Normalisation

A part of NLP pipeline for pre-processing text data, normalisation is about applying some linguistic models to tokens of text (Wiki et al., 2019). NLP is important in all the fields and especially in the field of corporate and cultural fields. It is also important for a wide range of people to have a working cognition of NLP within, for example, business information reasoning and web or app software alteration (Kaur et al., 2019) Moreover, the main goal of NLP understanding is to exploit the available resources like text corpora for semantic categorisation of texts.

Stemming and lemmatising are frequently considered as sibling processes and put under the same roof. The two procedures are connected and perform a similar function of reducing the variant words in the input text. The fundamental distinction between the processes lies in their outputs. The output product of stemming is "stem" and that of lemmatisation is "lemma." Stemming is extensively used as a pre-processing tool in the field of natural language processing, information retrieval and language modelling.

Now we will try to stem the words in the dataset and lemmatise them. Stemming is removing suffixes and prefixes and keeping just the root of each word (for example, Tables—Table), while lemmatising is getting the lemma of the word. The aim of lemmatisation is to reduce inflectional forms to a common base form (for example, Student—Stud).

3.4 Feature Extraction Algorithm

Feature extraction techniques remove the irrelevant features from the text documents and reduce the dimensionality of feature space. Feature extraction can be defined as a process of extracting a set of new features from the features set that is generated in the feature selection stage (Shah et al., 2016). To train the ML algorithms, these features are then created for vector creation. For our approach in this project, two types of vectors were used, which are CountVectorizer and TfidfVectorizer to, to extract the features from all the texts. CountVectorizer takes into account the frequency of features whereas TfidfVectorizer tries to determine the importance of a feature so that the classifier does not miss less frequent but important features (Eshan, 2017). In this chapter we used the function "get_features_names()" to see the most-used features we extracted from the text that will be used in classification; "fit_tranforms" is used to transform the training texts into features to use them in the classification; while the "transform" function is used to transform the testing texts into features to use them in testing our classifiers.

3.4.1 CountVectorizer

This feature extraction algorithm is commonly used in methods of document classification where the (frequency of) occurrence of each word is used as a feature for training a classifier. The CountVectorizer provides a simple way to both tokenise a collection of text documents and build a vocabulary of known words, but also to encode new documents using that vocabulary (Brownlee, 2019).

3.4.2 TfidfVectorizer (TF–IDF)

TfidfVectorizer (TF–IDF) stands for term frequency–inverse document frequency (also called tf–idf), and is a statistical method of finding weights of words (terms) in a document with regard to a list of documents (corpus), and it has two components: TF and IDF.TF (Term Frequency). In the simplest case, term frequency is just the number of times a word occurs in a document divided by the number of all terms in the document. And IDF (Inverse Document Frequency), using TF alone for calculating document

TABLE 3.5

Comparative Analysis of Feature Extraction Algorithm

Feature Extraction	Advantage	Disadvantage
CountVectorizer	– Provides an easy way to both tokenise a collection of text documents and build a vocabulary of known words, but also to encode new documents using that vocabulary [30]. – The encoded vectors can be used directly with machine learning algorithms [30].	– The CountVectorizer is more computationally expensive as compared to other algorithms [28].
TfidfVectorizer (TF–IDF)	– TF-IDF is an efficient and simple algorithm for matching words in a query to documents that are relevant to that query [33]. – Furthermore, encoding TF-IDF is straightforward, making it ideal for forming the basis for more complicated algorithms and query retrieval systems [33].	– TF-IDF does not make the jump to the relationship between words [33].

word weights, may not be optimal because it only operates within the scope of a single document.

Tf-idf Vectorizer is a concept which uses weight of the TD–IDF. When a term frequency is reached then we have a high-term frequency (TF) in the given document and low document frequency of the term in the whole collection. The inverse document frequency is a measure of how much information the word provides, i.e., if it is common or rare across all documents. It is the logarithmically scaled inverse fraction of the documents that contain the word (obtained by dividing the total number of documents by the number of documents containing the term, and then taking the logarithm of that quotient). Comparative analysis of feature extraction algorithm is shown in Table 3.5.

"CountVectorizer" and "TfidfVectorizer" functions take: "ngrame_range"= (1, 2), which means that we want one word and two words, whereas "max_features" is the number of features that we want, and it can be more or less. The unigram, bigram and trigram features are extracted from all the review text and vectorised using CountVectorizer and TfidfVectorizer to then train into the machine learning algorithms.

3.5 Supervised ML Classification

Supervised learning is where you have input variables (x) and an output variable (Y) and you can use an algorithm to learn the mapping function from the input to the output. Y = f(X): the goal is to approximate the mapping function so well that when you have new input data (x) you can predict the output variables (Y) for that data (Jolly and Gupta, 2018). This is supervised learning, because the process of an algorithm learning from the training dataset can be thought of as a teacher supervising the learning process, while unsupervised learning is where you only have input data (X) and no corresponding output variables. Six Supervised ML Classification models/algorithms that it have implemented this approach are listed:

A. Random Forest[*]
B. Logistic Regression[†]
C. Neural Network[‡]
D. SVM[§]
E. AdaBoost[¶]
F. Naïve Bayes[**]

3.6 Experiment Design

In addition, several combinations have been made among the features to find other features with which model/algorithms produced best results for prediction by means of the accuracy of classification for ML. Training the classifiers is done by the Train/Test Split. This test size is 20% of the dataset and the training set is the remaining 80% as shown in Figure 3.5.

In addition, an N-gram Model was used for extracting Unigrams, Bigrams and Trigrams in the dataset, as shown in Figure 3.6.

- Unigrams → are single words like → Cost
- Bigrams → are two words like → Work Great
- Trigrams → are three words like → Should highly recommend

Figure 3.6: In our approach to experiments/combinations, we try multiple times to get the best accurate results. In other words, for example in this

---

[*] Random Forests are an ensemble learning method for classification, regression and other tasks that operate by constructing a multitude of decision trees.
[†] Logistic Regression is a statistical model that in its basic form uses a logistic function to model a binary dependent variable, although many more complex extensions exist.
[‡] Neural Network is a series of algorithms that endeavors to recognise underlying relationships in a set of data through a process that mimics the way the human brain operates.
[§] SVM is a discriminative classifier formally defined by a separating hyperplane.
[¶] AdaBoost focuses on classification problems and aims to convert a set of weak classifiers into a strong one.
[**] Naïve Bayes is a conditional probability model that, given a problem instance to be classified, represented by a vector representing some n features (independent variables), assigns probabilities to this instance.

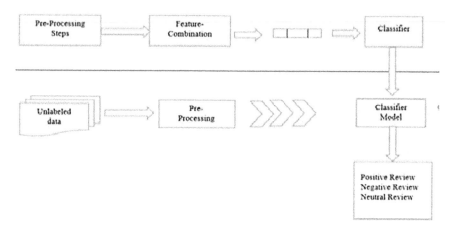

FIGURE 3.5
Train/Test Split in ML classification.

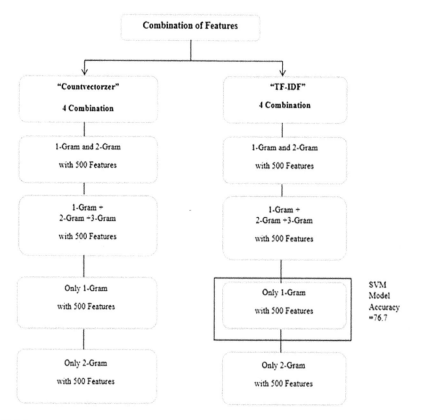

FIGURE 3.6
Combinations between features and grams experiments.

['abil', 'abl', 'absolut', 'access', 'account', 'activ', 'actual', 'ad', 'add', 'addict','addit','ago', 'allow', 'almost', 'alreadi', 'also', 'alway','amaz','amount','anim','annoy','anoth','answer','anyon','anyth','app','app great','appstore','app use', 'app work','app would','appl','applic','area','around','ask','avail','away','awesom', 'back','background','bad','base', 'basic','becom','best','better','big','bit','book','bore','bought','bug','build','busi','button','buy','call','came','camera', 'can','cannot','car','card', 'care', 'caus','certain','challeng','chang','charact','charg','chat','check','child','choos', 'clear','click','close','coin','color','come','commun','compani','complet','comput','connect','contact','content' ,'continu','control','cool','correct','cost','could','coupl','crash','creat','custom','custom service'].

FIGURE 3.7
Sample of unigram features for the dataset after being extracted from (A to C) with TfidfVectorizer algorithm.

['abl get', 'abl see', 'abl use','absolutlove','ad game', 'ad pop', 'also like', 'also love', 'also think','also would', 'amazapp','amount time','anoth app','anoththing','answerquestion','appabl','app actual','appallow','app also', 'app amaz', 'app anyon','app app','appawesom','app best','appcan','appcould','app crash' 'app develop','appdoesn','app don','app download','app easi', 'app even', 'app ever','appeveryth', 'app far', 'app find', 'app first','app get','appgive','appgo','appgood','appgot','appgreat','apphelp','apphope','apphowev','app iphon','app it','appkeep','applet','applike','applook','applot','applove','appmade','appmake','appmani','app much','appneed','appnever','appone','appopen','appperfect','appphone','appplea','apppretti','app purchas','app realli','app run','appsay','appsee','appseem','appshow','appsinc','app star','appstill','appstore','appsupport','apptake','appthank','appthink','apptime','apptri','appupdat', 'app use','appuser','appve','app want','appwell','appwork','appwould','app year','applwatch','ask question','awesomapp','backapp','backforth','bestapp','bestgame','boughtapp','bringback','bugfix','buyapp','can even', 'can

FIGURE 3.8
Sample of bigram features for the dataset after extracted from (A to C) with Count Vectorizer algorithm.

experiment, we try with 1,000, 2,000, 3,000 features but still the results are not very high in accuracy until it is found that 500 features are the best results. Then, we accredit 500 features in the rest of the steps for these experiments in this study. Figures 3.7 and 3.8 are the samples of features after having been extracted from our dataset, Figure 3.7 after using TfidfVectorizer and Figure 3.8 after using CountVectorizer, which includes alphabet characters from (A) to (C).

For each class in the ML classification models, we will produce a classification report, which include accuracy, precision, recall, f1-score and support, according to the following formulas (1–5) (Sokolava, 2009):

$$Accuracy_A = \frac{TP_A + TN_A}{TP_A + FN_A + FP_A + TN_A} \tag{1}$$

$$Precision_A = \frac{TP_A}{TP_A + FP_A} \tag{2}$$

$$Recall_A = \frac{TP_A}{TP_A + FN_A} \tag{3}$$

TABLE 3.6

Confusion Matrix for Binary Classification (Sokolava, 2009)

Data Class	Classified as pos	Classified as neg
Pos	True positive(tp)	False negative (fn)
Neg	False positive (fp)	True negative (tn)

$$F\_Measure_A = \frac{2.Precision_A.Recall_A}{Precision_A + Recall_A} \quad (4)$$

While the confusion matrix, represented by the following matrix in Table 3.6 is:

$$\begin{matrix} tp[fb] \\ fp \ tn \end{matrix} \quad (5)$$

3.7 Experimental Results and Analysis

The main results for our approach, the main/best three for combination/experiment, are:

(1) As it is clear that combinations between features and n-grams experiments bigram produced the best results accuracy as found in TD–IDF algorithm with the third combination (only one gram (unigram) with 500 features) with SVM model, the accuracy percentage was 76.57%. The following table shows the results (percentage accuracy) for all six ML algorithms within the best combination/experiments as shown in Tables 3.7 and 3.8.

TABLE 3.7

The Best First Results for All Models

Name of the Model	Accuracy Percentage %
Naïve Bayes	72.84
Random Forest	72.24
Logistic Regression	76.42
SVM	**76.57**
AdaBoost	73.2
Neural Network	72.73

TABLE 3.8

Result of the Confusion Matrix for the SVM Model (Best Combination/Experiment)

	Confusion Matrix		
	Pred: negative	**Pred: neutral**	**Pred: positive**
True: negative	893	157	251
True: neutral	1	6	4
True: positive	378	263	2547

['abil','abl','absolut','access','account','activ','actual','ad','add','addict','addit','advertis', 'ago','allow','almost','alreadi','also','alway','amaz', 'amount','anim','annoy','anoth','answer','anyon','anyth','app','appear','appl','applic','area','around','ask','avail','away','awesom','back', 'background','bad','base','basic','becom','begin','believ','best','better','big','bit','block','book','bore','bought','bug','build','busi','button','buy', 'calcul','call','came','camera','can','cannot','car','card','care','caus','certain','challeng','chang','charact','charg','chat','check','child','choic', 'choos','clear','click','close','coin','color','come','commun','compani','complaint','complet','comput','connect','constantli','contact','content', 'continu','control','cool','correct','cost','could','coupl','cours','crash','creat','current','custom','daili','data','date','day','deal','decid','definit', 'delet','design','develop','devic','didn','differ','difficult','disappoint','doesn','don','done','download','drive','earn','easi','easier','easili','edit', 'either','el','email','end','enjoy','enough','enter','error','especi','etc','even','event','ever','everi','everyon','everyth','exampl','excit','expect', 'experi','extra','extrem','face','facebook','fact','famili','far','fast','favorit','featur','feel','figur','file','final','find','fine','finish','first','five','fix', 'follow','forc','forward','found','free','friend','friendli','frustrat','full','fun','function','futur','game','gave','gener','get','give','glitch','go','goe', 'good','got','graphic','great','group','guess','guy','half','hand','happen','happi','hard','help','high','highli','hit','home','hope','hour','hous', 'howev','idea','import','improv','includ','info','inform','instal','instead','interest','interfac','io','ipad','iphon','isn','issu','it','item','job','keep', 'keyboard','kid','kind','know','languag','last','later','le','learn','least','leav','left','let','letter','level','life','light','like','limit','line','link','list','listen' ,'littl','live','ll','load','locat','lock','log','long','longer','look','lose','lost','lot','love','made','make','manag','mani','map','match','may','mayb', 'mean','meet','messag','might','mind','minut','miss','mobil','mode','money','month','move','much','multipl','music','must','name','need', 'network','never','new','next','nice','night','note','noth','notic','notif','number','offer','often','ok','old','one','onlin','open','option','order','other', 'otherwis','overal','page','paid','part','pas','password','past','pay','peopl','perfect','person','phone','photo','pick','pictur','place','plan','play', 'player','plea','plu','point','pop','possibl','post','power','practic','pretti','price','pro','probabl','problem','process','product','profil','program', 'progress','provid','purchas','put','puzzl','qualiti','question','quick','quickli','quit','rate','rather','re','read','real','realli','reason','receiv','recen, 'recommend','record','rememb','remov','report','request','requir','respons','review','right','room','run','said','save','say','schedul','school', 'screen','search','second','see','seem','seen','select','send','sent','servic','set','sever','share','show','sign','simpl','sinc','singl','site','slow','small',' social','someon','someth','sometim','son','soon','sound','spend','star','start','state','stay','still','stop','store','stori','student','stuff','subscript', 'suggest','super','support','sure','switch','system','take','talk','tap','team','tell','test','text','thank','that','there','they','thing','think','though', 'thought','three','time','today','took','tool','top','total','touch','track','tri','turn','two','type','understand','unless','unlock','updat','upgrad','use', 'useless','user','usual','ve','version','video','view','wait','want','wast','watch','way','websit','week','well','went','whole','wifi','win','wish', 'within','without','won','wonder','word','work','world','worth','would','write','wrong','year','yet','you'].

FIGURE 3.9
The result for 500 features in the best combination/experiment (TD–IDF algorithm with the third combination (0nly one gram – unigram)).

It is the best combination that has the highest score, and we will now list 500 features that were the result of this combination/experiment, as shown in Figure 3.9. The chart diagram for the six ML classifier for this combination/experiment is shown in Figure 3.10.

(2) The second-best results are obtained for combination of one (1-gram and 2-gram and 500 features) while also using TD-IDF algorithm for Logistic Regression model. The results (percentage accuracy) for all six models within this combination/experiment are shown in Tables 3.9 and 3.10.

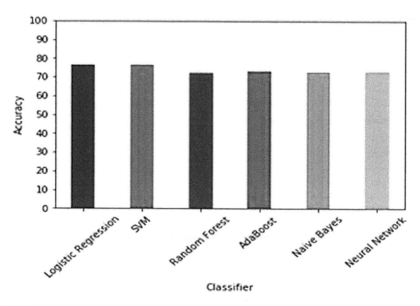

FIGURE 3.10

Chart of ML classifier in the best combination/experiment (TD–IDF algorithm with the third combination (only one gram – unigram)).

TABLE 3.9

The Second-Best First Results for All Models

Name of the Model	Accuracy Percentage %
Naïve Bayes	73.28
Random Forest	72.86
Logistic Regression	**76.22**
SVM	76.1
AdaBoost	73.42
Neural Network	73.11

TABLE 3.10

The Confusion Matrix for the Logistic Regression Model (Second-Best Combination/Experiment)

	Confusion		
	Matrix		
	Pred: negative	Pred: neutral	Pred: positive
True: negative	854	137	228
True: neutral	4	8	6
True: positive	414	281	2568

TABLE 3.11

The Third-Best Results for All Models

Name of the Model	Accuracy Percentage %
Naïve Bayes	73.24
Random Forest	71.84
Logistic Regression	**76.2**
SVM	**76.2**
AdaBoost	73.64
Neural Network	72.8

TABLE 3.12

The Confusion Matrix for the Logistic Regression Model (Third-Best Combination/Experiment)

	Confusion Matrix		
	Pred: negative	Pred: neutral	Pred: positive
True: negative	854	136	229
True: neutral	4	8	6
True: positive	414	282	25687

TABLE 3.13

The Confusion Matrix for the SVM Model (Third-Best Combination/Experiment)

	Confusion Matrix		
	Pred: negative	Pred: neutral	Pred: positive
True: negative	876	153	252
True: neutral	2	7	4
True: positive	394	266	2546

The third best results for this combination/experiment was combination two (1-gram and 2-gram and 3-gram, and 500 features) with also TF–IDF algorithm for both SVM and Logistic Regression. The results (percentage accuracy) for all six models within this combination/experiment are shown Tables 3.11, 3.12 and 3.13.

These experiments (the third one) show that these algorithms (SVM, Logistic Regression) topped the first ranks, while these algorithms (Random Forest, Neural Network) were the least performing at the level of those experiments in general. Moreover, it was also found that in TF-IDF Vectorizer the highest results were achieved from their counterpart Count-Vectorizer. 500 Features either in (uni-gram or bi-gram or tri-grams) was the optimal number for the 22,496 text reviews in our dataset; this number we achieved after cleaning the text (Jolly & Gupta, 2019) and performing either text normalization and text processing (NLP). Furthermore, our experiment shows that unigram might be able to give better results than bigram and trigram and we did not find any results when we chose trigram and there did not appear any results in any of the three linked words that reflected trigram in the N-gram model. At the end of this section we will present two graphs for both of the feature extractions algorithms (CountVectorizer and TF-IDF) that summarise all of the results of ML classifier algorithms combined with the two mains of the feature extraction algorithms for each combination/experiment. This is followed by the top 20 important words, positive, negative and neutral, that have been inferred from this study as illustrated in Figures 3.11, 3.12 and 3.13.

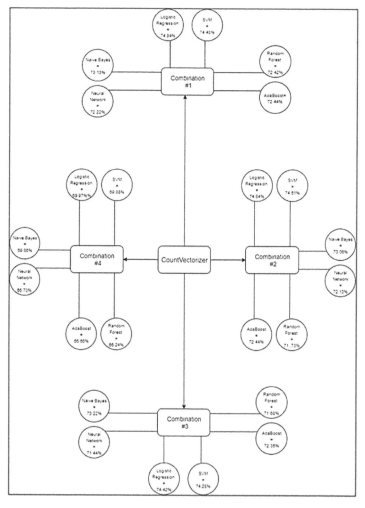

FIGURE 3.11
Results for all combination/experiment in CountVectorizer.

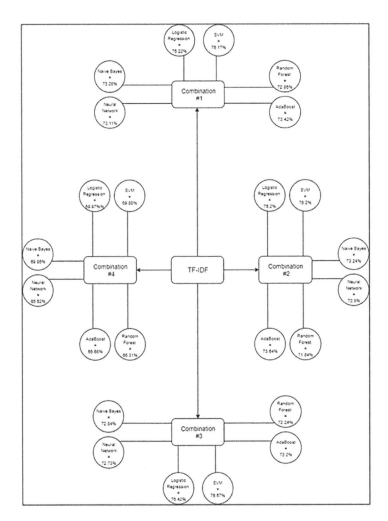

FIGURE 3.12
Results for all combination/experiment in TF–IDF.

3.8 Recommendation and Future Work

First of all, we recommend using SVM and Logistic Regression classification algorithms when we need to handle the text data for reviews.. TF–IDF algorithms achieved higher results than CountVectorizer for all four combinations of experiment, so we recommended using them as a major algorithm for extraction texts in general.

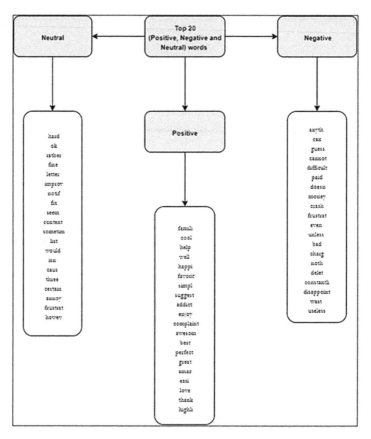

FIGURE 3.13
Top 20 words, positive, negative and neutral, that were concluded in this study.

As future work, we will extend the dataset to include more reviews from other categories such as News, Shopping, Finance etc., to know more about the main features that are included in other categories. Moreover, as a team, we plan to improve the applied classification model to predict the type of user reviews, such as feature request, bug report etc. Besides that, the category of the application will also belong to negative or positive or neutral review and may also belong to business category, for example. In this way, we will improve decision makers for multiple domains by answering this important question "what is the major features that the people try to seek for each category?".

As an extension of this study, we are planning to build adaptive prediction for text reviews for the mobile application domain as a system to allow the app developer, by using this proposed system, to know the percentage of customer satisfaction and dissatisfaction.

3.9 Conclusion

In this research project, we collected and studied 22,496 text reviews from iTunes Apple Store. We extended our analysis in our approach by integrating two feature extraction algorithms (CountVectorizer and TF–IDF) and sentiment analysis with six ML classification algorithms (Random Forest, LR, Neural Network, SVM, AdaBoost, NB) and we made several combinations among the features to find any of the features with which ML algorithms produce the best text review prediction of sentiment analysis after we had applied feature extraction. The best results accuracy was found in TD–IDF algorithm with one gram with 500 features with Support Vector Machine (SVM) ML algorithm, and the accuracy percentage was 76.57%. Finally, these project results using our dataset showed the most common aspects of features that developers need to consider when developing or updating their apps.

References

Agrawal, R. and Batra, M., 2013, 'A detailed study on text mining techniques', *International Journal of Soft Computing and Engineering*, vol. 2, no. 6, pp. 118–121.

Agrawal, R. and Gupta, N. (eds.), 2018, *Extracting knowledge from opinion mining*, IGI Global: Hershey, PA.

Aiquwayfili, N., Alkomi, N., AlZakari, N., et al., 2013, 'Towards classifying applications in mobile phone markets: the case of religious apps', in *2013 international conference on current trends in information technology (CTIT)*, pp. 177–180.

Ali, M., Joorabchi, M. E. and Mesbah, A., 2017, "Same app, different app stores: a comparative study', in *2017 IEEE/ACM 4th international conference on mobile software engineering and systems (MOBILESoft)*, Buenos Aires, Argentina, pp. 79–90.

Brownlee, J., 2019, 07 August, 'How to prepare text data for machine learning with scikit-learn', *Machine Learning Mastery*. Available at https://machinelearningmastery.com/prepare-text-data-machine-learning-scikit-learn/ (accessed May 26, 2019).

Brownlee, J., 2019, 12 August, 'Supervised and unsupervised machine learning algorithms', *Machine Learning Mastery*. Available at https://machinelearningmastery.com/supervised-and-unsupervised-machine-learning-algorithms/ (accessed October 24, 2019).

Daniel, G., Lourenço, A., Hugo, L.-F., Miguel, R.-J. and Florentino, F.-R., 2014, 'Web scraping technologies in an API world', *Briefings in Bioinformatics*, vol. 15, no. 5, pp. 788–797.

Dhamija, V., 2019, 09 October, 'CountVectorizer: hashingTF', *Medium*. Available at https://towardsdatascience.com/countvectorizer-hashingtf-e66f169e2d4e (accessed November 08, 2019).

Eshan, S. C., 2017, 'An application of machine learning to detect abusive Bengali text', In *2017 20th International Conference of Computer and Information Technology (ICCIT)*, pp. 1–6.

Fu, B., Lin, J., Li, L., Faloutsos, C., Hong, J. and Sadeh, N., 2013, 'Why people hate your app: making sense of user feedback in a mobile app store', in *Proceedings of the international conference on knowledge discovery and data mining (KDD)*, pp. 1276–1284.

Gheorghe, M., Mihai, F. and Dârdal, M., 2018, 'Modern techniques of web scraping for data scientists', *Romanian Journal of Human-Computer Interaction*, vol. 11, no. 1, p. 2018.

Guo, T, Guo, B., Ouyangyi, Y. and Zhiwen, Y., 2018, 'Mining and analyzing user feedback from app reviews: an econometric approach', in *2018 IEEE smartworld, ubiquitous intelligence & computing, advanced & trusted computing, scalable computing & communications, cloud & big data computing, Internet of people and smart city innovation (SmartWorld/SCALCOM/UIC/ATC/CBDCom/IOP/SCI)*, Guangzhou, China, pp. 841–848.

Gupta, N. and Verma, S., 2019, 'Tools of opinion mining', in *Extracting knowledge from opinion mining*, IGI Global: Hershey, PA, pp. 268–296.

Guzman, E. and Maalej, W., 2014, 'How do users like this feature? A fine grained sentiment analysis of app reviews', in *Proceedings of the 22nd IEEE international requirements engineering conference (RE)*, pp. 153–162.

Guzman, E., El-Haliby, M. and Bruegge, B., 2015, 'Ensemble methods for app review classification: an approach for software evolution (N)', in *Proceedings of 30th IEEE/ACM international conference on automated software engineering*, pp. 771–776.

Harman, M., Jia, Y. and Zhang, Y., 2012, June. App store mining and analysis: MSR for app stores. In *2012 9th IEEE working conference on mining software repositories (MSR)* (pp. 108-111). IEEE.

Jha, N. and Mahmoud, A., 2018, 'Using frame semantics for classifying and summarizing application store reviews', *Empirical Software Engineering*, pp. 1–34. Available at http://link.springer.com/10.1007/s10664-018-9605-x.

Jolly, S and Agrawal, R., 2019, 'A broad coverage of corpus for understanding translation divergences', *International Journal of Innovative Technology and Exploring Engineering (IJITEE)*, vol. 8, no, 8S2, pp. 2278–3075.

Jolly, S. and Gupta, N., 2019, 'Handling mislaid/missing data to attain data trait', *International Journal of Innovative Technology and Exploring Engineering*, vol. 8, no. 12, pp. 4308–4311.

Jolly, S. and Gupta, N., 2018, 'Extemporizing the data trait', *International Journal of Engineering Trends and Technology*, vol. 58, no. 2, pp. 100–103.

Jolly, S. K. and Agrawal, R., 2019, 'Anatomizing lexicon with natural language Tokenizer Toolkit 3', in *Extracting knowledge from opinion mining*, IGI Global: Hershey, PA, pp. 232–266.

Kalra, V. and Agrawal, R., 2019, 'Challenges of text analytics in opinion mining', in *Extracting knowledge from opinion mining*, IGI Global: Hershey, PA, pp. 268–282.

Kaur, S and Agrawal, R., 2018, 'A detailed analysis of ore NLP for information extraction', *International Journal of Machine Learning and Networked Collaborative Engineering*, vol. 1, no. 01, pp. 33–47.

Kim, J., Park, Y., Kim, C. and Lee, H., 2014, 'Mobile application service networks: Apple's app store', *Service Business*, vol. 8, no. 1, pp. 1–27.

Kimbler, K., 2010, 'App store strategies for service providers', in *2010 14th international conference on intelligence in next generation networks (ICIN)*.

Lohmann, S., Heimerl, F., Bopp, F., Burch, M. and Ertl, T., 2015, 'ConcentriCloud : word cloud visualization for multiple text documents', in *2015 19th international conference information visualisation*, pp. 114–120.

Maalej, W. and Nabil, H., 2015, 'Bug report feature request or simply praise? on automatically classifying app reviews', in *Proceedings of IEEE 23rd international requirements engineering conference*, pp. 116–125.

Maalej, W., Kurtanović, Z., Nabil, H. and Stanik, C., 2016, 'On the automatic classification of app reviews', *Requirements Engineering*, vol. 21, no. 3, pp. 311–331.

McIlroy, S., Ali, N., Khalid, H. and Hassan, A. E., 2016, 'Analyzing and automatically labelling the types of user issues that are raised in mobile app reviews', *Empirical Software Engineering*, vol. 21, no. 3, pp. 1067–1106.

Pagano, D. and Maalej, W., 2013, 'User feedback in the appstore: an empirical study', in *Proceedings of the international conference on requirements engineering - RE '13*, pp. 125–134.

Panichella, S., Sorbo, A. D., Guzman, E., Visaggio, A., Canfora, G. and Gall, H., 2015, 'How can I improve my app? Classifying user reviews for software maintenance and evolution', in *Proceedings of 31st IEEE international conference software maintenance evolution*, pp. 281–290.

Powers, D., 2011, 'Evaluation: from precision recall and f-measure to roc informednessmarkedness and correlation', *Journal of Machine Learning Technologies*, vol. 2, no. 1, pp. 37–63.

Ramos, J., 2003, 'Using tf-idf to determine word relevance in document queries', in *Proceedings of the first instructional conference on machine learning*.

Shah, F. P., 2016, 'A review on feature selection and feature extraction for text classification," in *2016 international conference on wireless communications, signal processing and networking*, pp. 2264–2268.

Singh, J. and Gupta, V., 2016, 'A systematic review of text stemming techniques', *Artificial Intelligence Review*, vol. 48, no. 2, pp. 1–61.

Sokolova, M. and Lapalme, G., 2009, 'A systematic analysis of performance measures for classification tasks', *Information Processing and Management*, vol. 45, no. 4, pp. 427–437.

Vakulenko, S., Müller, O. and Brocke, J. V., 2014, 'Enriching iTunes app store categories via topic modeling', in *Proceedings of the international conference on information systems*, pp. 1–11.

Yang, K. F., Yang, H. W., Chang, W. Y. and Chien, H. K., 2017, 'The effect of service quality among customer satisfaction, brand loyalty and brand image', in *2017 IEEE international conference on industrial engineering and engineering management (IEEM)*, Singapore, pp. 2286–2290.

Zebende, G. F., Silva, M. F. and Filho, A. M., 2013, 'DCCA cross-correlation coefficient differentiation: theoretical and practical approaches', *Physica A*, vol. 392, no. 8, pp. 1756–1761.

4

Machine Learning for Biomedical and Health Informatics

Sanjukta Bhattacharya and Chinmay Chakraborty

CONTENTS

4.1 Introduction

Nowadays, ML-based applications are used in several industries such as finance, travel, media, healthcare, oil and gas, retail, social media, marketing and sales, data security, transportation, etc. Several real-life applications are built and developed with the help of ML such as virtual personal assistants, customer relationship management, video surveillance, human resource segment, email spam and malware filtering, product recommendation, business intelligence and analytical management, online customer support, computer vision, search engine result refining, self-driving cars, online fraud detection, virtual assistants. In recent decades the healthcare sector has been constantly overwhelmed with various technological influences. ML is one of them, and has a huge impact on medical science. ML that is taken as a development area of AI is also combined with computational biology, statistics, statistical physics, information, electrical engineering and applied mathematics to increase its remarkable learning capability to solve any real-time complex problems very quickly with optimum accuracy (Clifton and Gibbons 2012).

The objective of ML is to create algorithms that have the capability to automatically learn and then improve themselves so that these experienced and improved algorithms can be used for future predictions. ML is a broad field that has recently been growing very quickly and is essential for extracting useful features or patterns to solve complicated tasks with proper accuracy, which includes classifying, decision support, anomalies detection, forecasting, sentimental analysis and ranking. ML has the ability to extract or discover the relevant patterns or features in a vast amount of data that are not properly accessible and not properly visible to human beings (Holzinger 2016). Generally, the algorithms of ML are categorised into three subcategories, which are defined as supervised learning, unsupervised learning and reinforcement learning. In the case of supervised learning, the input and output features, or patterns pair, are given as a training set, and then the sample or input features are given to the systems as a test set and the system predicts the accurate output based on the previous learning experience from the training set. In the case of unsupervised learning, only the input features are given as a training set where the target-based output is not associated with any input feature. These input features are used to extract suitable, meaningful patterns. Over the last few decades, different ML algorithms have been developed, which are used in different applications to get tremendous accuracy, and that is why ML is truly well-accepted in several important and crucial sectors (Chen et al. 2005). Since 1940, researchers have created intelligent, knowledgeable computers with learning capability, and which are mainly used for applications such as troubleshooting problems, any decision-making procedure within the business environment, diagnostic processes of any medical issues, but these systems are not capable of extracting knowledge automatically. So, as a result, a huge amount of labor and time is required for the completion of these systems. The same thing happens in the medical sector also, because this conventional method is not capable of handling a huge amount of complex data with high variety, high volume and high velocity. Now ML is being introduced that comprises a set of algorithms that are very capable of extracting knowledge from sources or examples automatically. This ML approach is responsible for sharing, searching and capturing data, and analyzing and storing different types of data from the medical field. In the case of the medical sector, ML is used to guarantee reductions in the cost of patient care, to help physicians and doctors create personalised opinions, to improve the effectiveness and timeliness of treatments and to get personalised and accurate medicine to patients in time (Ravì et al. 2017).

Health informatics is an area where data is gathered with the highest priority. Health-related data are collected from various sources: such as the sensor-related data that can be obtained from medical monitors and instruments automatically; clinical data from the hospital or nursing home where the patient may be admitted for a period of time; data that can be collected from practitioner interactions with patients; the social-related dataset that can be collected from patients' home and obtained mainly from patients who are suffering from critical diseases for long periods of time; and the genomic data that can be achieved from different medical tests or notes and which can be obtained from medical consultants. These collections of data are extremely large in volume and very complex in nature (Clifton and Gibbons 2012). The aim of health informatics is to upgrade the quality of patient care quality and to enhance the output of the healthcare segment. Nowadays, the worldwide accessibility and the tremendous growth of medical data are very much responsible for various activities where the fields of data science truly combine. ML performs a vital role in the case of any information or image that is used in the medical field. To properly and accurately understand health informatics in the biomedical field, the medical dataset, which is very large, diverse and complex in nature, must organise, manage and manipulate clinical-based knowledge (Fang et al. 2016).

The biomedical field has now attracted more attention and become a priority because it is directly related to issues of human health, which is facing new challenges every day. A large volume of data related to research in the biomedical field is constantly generated through various biomedical, molecular and genomic applications and techniques such as identification of protein, medical records of patients, medical imaging, genome sequencing. The interpretation rate of this large volume of data is much faster than the accumulation rate of those data. This huge amount of data is, first, efficiently organised and then properly analyzed, which is very useful to different applications in a well-organised and systematic way (Chen et al. 2005). Various newly invented information methods and computational methods are required to properly organise this large volume of biomedical-related data, and after that to efficiently extract useful patterns or features to accurately predict the solution of any complex or critical task. Moreover, in the case of the biomedical sector the data sets are non-standardised, incomplete, poorly structured, uncertain in nature, such as dirty data, missing data, unwanted data and noisy data, and also the dimensions of maximum data sets are higher than dimension 3 so it is quite impossible for the human professional to manually analysis the pattern accurately (Holzinger 2016).

ML consists of various methods to correctly and efficiently retrieve and share biomedical information. ML is applied to effectively discover the features or patterns from different biological areas such as molecular or genomic fields, drug discovery, the patient-related care segment, and then to analyze them with optimum accuracy. Using ML, the entities related to the biomedical area such as proteins, the names of drug, diseases and genes are automatically discovered, in order to accurately develop the pathways of a gene or provide suitable mapping within the existing ontologies of the biomedical field (Chen 2005). ML is defined as an essential platform that is used in biological and biomedical datasets for analyzing and extracting the required patterns. This tool is also capable of identifying the relationships hidden in healthcare-related data and also the critical structure of the data.

4.2 Overview of Machine Learning Applications

The data analysis procedure through ML is used to automate the model of analytical building so that systems are used to learn from the given example or data, properly identify the accurate patterns from the previous experience and then make suitable decisions where human intervention is required as little as possible. Nowadays, most of the industries that are continually dealing with a large amount of data have recognised the essence, value and importance of ML, which is very helpful for organisations to work efficiently, and collect or gather knowledge to gain worldwide advantage. Different applications are created to serve various types of industrial requirement such as financial services, the healthcare sector, the oil and gas industry, the government sector, transportation services, the sales and marketing sector, the legal industry, automatic number recognition, digital libraries and institutional repositories, handwriting recognition, optical music recognition, the banking sector, invoice imaging, bio-surveillance, accelerating empirical sciences, speech recognition, robot control, computer vision. Among all industries, the healthcare sector has expanded enormously based on this ML, where the highest level of accuracy is always needed in order to predict known or unknown diseases or illnesses. Several applications have been developed that make diagnostic procedures effective and accurate, so that human beings benefit from treatment processes or the diagnosis of any procedure critical diseases.

1) Supervised and unsupervised ML is used for the development and discovery of drugs, especially in the preliminary phase, which starts with early-stage screening, and is also used in precision medicine for better understanding the mechanisms of diseases as well as the better route of treatment for those diseases. Various companies

(Microsoft and MIT) have developed efficient algorithms of ML to improve the diagnostic procedure for diseases, especially for leukemia or cancer and also to enhance capabilities of fast and appropriate predictions.

2) In the case of personalised treatment, with the help of ML a large volume of data, collected from patients through healthcare and IoT devices, is interpreted in a well-organised and systematic way and then the interpreted data are used to accurately predict the state of diseases or to recommend suitable treatment. Several companies like IBM are providing various healthcare solutions that are developed through ML and AI to increase accuracy levels.

3) Using different ML applications, the field of medical surgery is growing quickly, where robots are used to replace human surgeons. This has many advantages, for example reducing the challenges usually faced by human beings, such as hands trembling in an operating theater, or allowing operations to be performed in smaller spaces and in fine detail. Robotic surgery is used to measure accuracy to a higher degree and also identify particular organs or parts within the human body.

4) ML handles the enormous amount of medical data from different sources. Data-handling, which is done efficiently through ML, is a very important stage in the accurate analysis of medical images for critical diseases such as Asperger's syndrome or Parkinson's disease.

5) ML-based medical imaging is very important in terms of diagnosis and prediction of disease. Different companies have developed various applications that are used to appropriately analyse the medical images coming from medical devices and to predict the optimum level of accuracy for the detection of diseases or to recommend suitable treatment.

6) Automation technology integrated into different healthcare solutions that is based on ML makes a significant improvement in various medical sectors, specifically in the field of surgery.

7) Clinical trials, such as identifying the absolute candidate category based on some factors, benefit from the use of ML algorithms. Using these algorithms, clinical trials, where medical experts can access a complete range of medical data, decreases cost, as well as the time that is required to conduct medically based experiments that can be performed quickly and more efficiently and with less memory storage.

8) It is a significant challenge to properly maintain the massive volume of medical data that are collected from different sources. ML plays an important role in storing the acquired data electronically, so that it can be interpreted later on in order to propose treatment plans or to efficiently explain a patient's condition.

9) ML, along with AI, not only performs an appreciable role in various medical sectors but also plays a very lucrative role in the prediction of outbreaks of diseases, epidemics or malaria in any Indian state. This prediction of outbreak of disease generally depends on climate, such as temperature or the amount of rainfall on a specific day in a particular area.

10) Radiation-based treatments benefit highly from the use of ML. Before applying any radiation treatments such as radiotherapy, accurately detecting what is infected tissue and what is normal tissue is of primary importance, because the probability of damaging normal and healthy cells is very high.

Figure 4.1 describes different domains or segments of today's socioeconomic life that are truly affected by and benefit from accurate predictions or efficient identification of living or non-living entities using ML techniques.

In financial services, ML is mainly used to identify vital insight into the data and, of course, prevent forgery. This crucial insight is used for the identification of investment-related opportunities and also to help investors make proper decisions for trading. The healthcare industry is also very much influenced by ML, where the sensors and other devices are used to detect and observe the health of the patient. This technology is also very beneficial

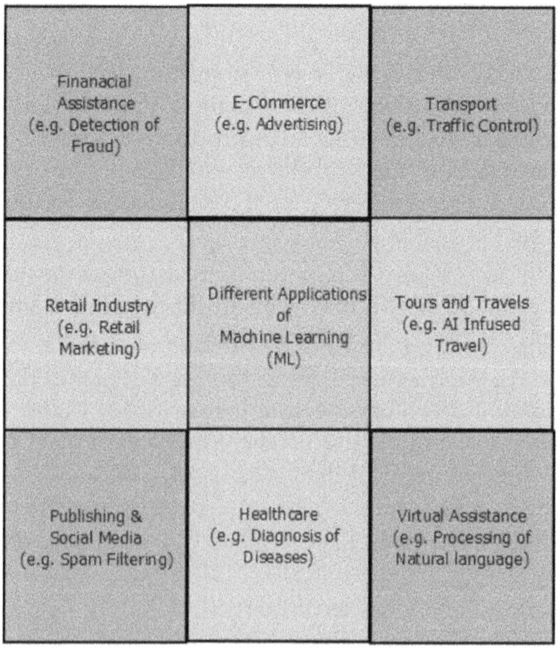

FIGURE 4.1
Different ML application areas.

for data analysis by medical specialists and also for improving diagnosis or treatment. The oil and gas industry benefits from ML, which is used to analyze minerals within the ground, to find new sources of energy and also to predict the failure of any refinery-based sensors. In the case of the government sector, this ever-developing technology is used for safety purposes for public services and other utilities that are important for maintaining the socio-economic structure. The transportation sectors also benefit from ML. Public transportation, traffic control, vehicle maintenance and road safety are the major areas of challenge finely handled by ML. In the industry of sales and marketing, the ML is applied to solve problems such as analyzing buyers' purchasing histories or for the purpose of promoting any item, along with the ability to capture important data, which are analyzed further for retail marketing campaigns. The legal industry is also taking advantage of ML by using optical character recognition (OCR) technology, used for the digitisation of documents, which is very helpful for the legal expert who wants to quickly search for documents in a huge database.

ML is used to recognise or identify the registration plates of vehicles by using a mass surveillance procedure where the recognition of optical characters in the image is very important. This surveillance technique is used by several police forces or within electronically controlled toll tax systems. Institutional repositories are vital collections of data that are digitally maintained using ML. These huge collections of digital data are very carefully collected and attentively preserved and are used with significant institutional impact throughout the world. This technology is also used to properly maintain monographs, theses, proceedings, journal articles and research data. It is used to recognise handwritten characters very efficiently where the input is from sources such as a touch screen, paper documents, photographs and other devices and which then needs to be correctly interpreted. Optical music recognition (OMR) is developed to recognise and edit printed sheets into a playable format. This OMR is also used to process different levels of music, to distinguish between various music notations and also to digitise large volume of musical data. The banking sector also uses the OCR system that is developed through ML for the efficient processing of cheques and other banking activities without human intervention, in order to reduce waiting time. Invoice imaging is very much influenced by ML technology and that is why several applications have been developed to better serve invoice imaging (Libbrecht and Noble 2015). Bio-surveillance is a vital area where ML helps to track and detect diseases. Current work involves adding a rich set of additional data, such as retail purchases of over-the-counter medicines, to increase the information flow into the system, further increasing the need for automated learning methods, given this more complex data set. ML is very beneficial for learning the purpose of the gene-related model within the cell body from the data which is very high in terms of throughput in nature, for characterisation of the complicated patterns of the activation of brain and also for the discovering purpose of some unusual objects from a huge collection of data. Speech recognition is another field where ML performs an essential role in recognising different types of speech received from different types of resources (Tom 2006).

4.3 Impact of Machine Learning in Healthcare

ML has an important impact on image informatics and the medical imaging field. An enormous number of challenges exist in this field, which include heterogeneous and diverse inputs, the accuracy of a pattern of a key that is hidden through huge discrete variation, insufficient instances versus the features that are high dimensional in nature, and occasionally an undisclosed mechanism that is underlying a disease. Nowadays, the field of healthcare is experienced in the fast growth of data. The large volume and variation of complex medical data are always required to represent the required knowledge through the continuous learning process for a proper understanding of the area of health informatics as it is the real data. This huge volume of medical data is used to obtain accurate clinically based knowledge that can be used to take action to resolve a problem. In the early days, it was not easy to handle the complex, nature-based data with high volume, high velocity and high variety but since the invention of ML-based methods, it is much easier to handle the vast amount of data with optimum accuracy in a minimum period of time.

There are big changes and enormous challenges in ML-based medical imaging and because of this more and more people are now dedicating the direction of their research to the medical field along with ML. System-based modeling and health or biomedical informatics, which is inspired by ML, cover various fields at the points of intersection of biology, computational science and medicine. Many models that are mechanistic in nature, different methods based on the concept of AI and various simulations have been developed to accurately describe various observed medical data and biological occurrences, and then to derive various new organic or biologically based insights and, lastly, to translate the impacts of various scientific and newly invented discoveries for patient care and human health. Preprocessing, which include techniques such as integration, transformation, cleaning and reduction, etc. is the initial step in the pipeline of Machine Learning, which is used to resolve problems with data, such as missing data, inconsistent characteristics of data from the real world, noisy data and also to improve the quality of data.

Modeling is the second stage that is used for the construction of different data models such as probabilistic models and statistical models. These models are effectively used for the detection of patterns or features in the existing data and the detected patterns are then used again to predict future data. The next stage is an evaluation that performs the last or final phase where the complexity, technical correctness and the efficiency of the model are properly established. Validation and verification processes are performed to check the ability of an accurate representation of a model in the observed system and then to check that the model construction and data manipulation are experienced with technical correctness or not. After that, medical coding is required, where the data are systematically classified into proper alphanumeric codes for accurate and efficient identification of medications, diagnosis, laboratory tests, procedures and additional clinical properties (Clifton et al. 2015). Various applications that include clinical informatics, image analysis, bioinformatics, immune engineering, system biology, computational neuroscience, precision medicine have been developed in combination with other departments so that various sectors in the medical field derive benefit. At present, the health informatics systems that have been developed based on ML include a vast range of different technologies such as different sensors that are used to collect data from patients, different signal processing techniques in biomedical fields that are used for the conditioning of high-resolution-based data and big data approaches that are used to fuse the data acquired from various segments such as categorical data, which are collected from electronic health records and time-series data from different sensors.

The field of healthcare is now fully developed with ML, which is extensively used to finely manage and improve the medical system. With the advancement of ML, the overall cost of healthcare sectors have been surprisingly reduced, which is very beneficial to the general population. This also means that for patients at high risk, there is accurate prediction of disease based on ML. Different types of ML-based predictive model have been developed for the correct diagnosis and treatment of many diseases. This model is also used for personalised treatment and care such as for cancer treatment, where genomic DNA sequence is used and clinical operations, where analysis of unstructured and historical data of the patients are required to get more successful treatments. Public health is protected because of time detection, and after that proper prevention of any infectious diseases is possible very quickly and efficiently. Genomic analysis is one of the prime factors in recent healthcare where effective and efficient gene sequencing techniques are developed to capture the DNA sequence and also to perform the proper genome-based association for any critical disease/microbiome investigations of human beings. Continuous monitoring process for aged persons, which are carried out either in hospital or in-home, can be safely and successfully done through different wearable devices; miscellaneous issues such as fraud detection and security problems can also be solved with the help of ML (Feldman et al. 2017). The current development in health informatics in the biomedical field are based on ML and have influenced the clinical treatment procedures of patients very effectively (Clifton et al. 2015).

4.4 Recent Trends in Machine Learning in the Biomedical Field

Today different industries are benefited through the extensive use of ML, which is used not only to make the daily lives of human beings much more easier way but also to invent numerous real-life innovative applications to the benefit of mankind. With the help of ML, appreciable advances have been made such as autonomous vehicles, speech recognition, health informatics, recommender systems, etc. Computer vision is one of the important branches where applications are widely employed to get best performance. There are several problems that exist in computer vision such as image identification or recognition, automatic analysis of documents, speech recognition, detection of objects, medical image processing, the impact on genetics and genomics, face detection, etc. Healthcare is one of the prime areas where the algorithms of ML are well implemented and perfectly suited to create accurate outcomes. ML health informatics creates several business opportunities as well as employment for graduates and is also the latest important challenge for research. Various emerging ML-based applications are created in the medical devices sector and professionals are constantly trying to improve the technology. According to a recently published report, in the United States medical devices sector, manufacturers and researchers are trying to integrate accuracy with the help of automation. The latest emerging applications in healthcare have three prime divisions such as: medical imaging, where the ML platform is used to improve clinical outcomes and the clarity of images taken by medical scanning devices by reducing radiation exposure; managing chronic diseases, where ML is used to constantly monitor patients with the help of sensor technology; and to deliver treatment automatically with the help of connected mobile-based applications. Multiple medical device companies have now merged with ML to improve treatment, discover new drugs for chronic diseases and also to provide an optimum level of patient care.

Medtronic is one of the companies whose aim is to help healthcare in various ways. One of the main objectives of this organisation is to control diabetes more efficiently in order to create better lifestyles. With the collaboration of IBM Watson, Medtronic has also developed a mobile assistant application that is completely personalised, named Sugar IQ. This application has some specific features such as glycemic assistance, where therapy-based actions or diet are strictly monitored. Every meal is based on insights that uniquely specify the impact of the specific food on glucose levels. The application exposes behaviors that are related to patients' glucose patterns, sends beneficial personalised messages and continuously monitors glucose levels in the blood by the data-analyzing process generated through insulin pumps and Medtronic glucose sensors. Following this, another application, named the MiniMed 670G system, was developed by Medtronic, trained on ML algorithms in such a way that it is used to adjust and stage the delivery of baseline insulin. GE Healthcare collaborated with NVIDIA to develop various medical devices with the help of ML and it has captured the global market with its 5-lakh medical imaging device. These collaborations apply ML to improve the accuracy and speed of computerised tomography (CT) scans. Philips Healthcare has also developed ML-based wearable technology for clinical settings. This organisation developed a patient motoring system known as the IntelliVueGuardian Solution to timely and accurately predict the crisis moment of a patient and to provide proper medication on time. ML-based navigation of clinical records has been invented, where the patient's medical data are efficiently searched and then correctly analyzed for further use. Robin Healthcare has developed an ML-based virtual assistant that is used to store and document all clinical information in electronic health records (EHR), which are generated from real-time dialogues between patient and medical expert. In addition, through voice or text interactions virtual assistants are rapidly being used in the healthcare sector. Nuance Company has developed a virtual assistant based on ML, named Dragon Medical Virtual Assistant, that is used to automatically help document all clinical information. The MedWhat company developed another ML-based virtual assistant that is used to instantly respond to the health and medical questions from both medical experts and patients. The ML-based automation system is widely used to reduce the possibility of surgeon fatigue and the time of the surgical process. In the event of any surgical procedure, ML algorithms are used to compute features such as depth perception, completion time, curvature, speed, path length, smoothness, etc. ML is used to develop surgical skills and also used to improve the materials that are used for surgical robotic management.

Figure 4.2 shows various phases of ML capable of proper diagnosis of diseases. Apart from these systems, ML along with AI and computer vision is used to develop techniques through which patients' data are analyzed and interpreted accurately to diagnose disease at a very early stage and then to prescribe appropriate medication to improve treatment. For some of the critical diseases, such as cancer, detection at an early stage is also possible with the help of ML. Automated diagnosis and audit systems are used to decrease prescription-based errors and also discover the precise disease of the patient. Pregnancy management, where the prime factor is constantly monitoring the pregnant woman, is very much benefitted by the use of ML. Patient triaging solutions have been invented that are used to scan incoming cases for various clinical conditions, determine their priority and subsequently send them to the most appropriate medical expert within the

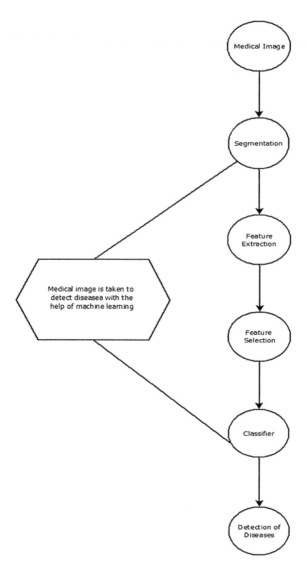

FIGURE 4.2
Role of ML in biomedical applications.

network. These technologies are not only beneficial for patient care but are also used to reduce medical and hospital costs. Early detection of cancer cell recognition and the accurate treatment procedures for different types of cancer in various parts of the human body can be initiated within a minimum amount of time. Multiple critical drug discoveries are another recent advance in the healthcare field. Gene analysis and different genetic diseases can be predicted and resolved at the root level with the help of updated medication and cost-effective procedures.

4.5 Different Machine Learning Techniques for Healthcare Sectors

ML is the combination of statistics and computer science where the learning process is achieved with the help of several computing algorithms and a collection of data is categorised into four main groups such as supervised learning, unsupervised learning, reinforcement learning and semi-supervised learning (Rahul 2015). ML is applicable to those medical datasets that are high-dimensional, large-scale and multidimensional in nature. ML is comprised of various algorithms that are developed in such a way they can improve themselves when new data arrive. Two sets of data, such as a training data set and a test data set, are required in ML applications where the training set of data are used to train the algorithms, after which the test data set is fed to the algorithm so that the algorithm makes a logic/prediction based on the given new data, which are then analyzed for accuracy. This accuracy is very important for accepting the purpose of the ML algorithm. Until and unless the accuracy is not well accepted, the training process of algorithm will not stop. Algorithms that fall into the category of supervised learning are K-Nearest Neighbor (KNN) and Support Vector Machine (SVM). Unsupervised learning includes algorithms such as K-Means clustering and Hierarchical clustering. Another advance categorisation is known as deep learning (DL) where the artificial neural network (ANN) is an important concept. There are certain steps that are usually involved in ML when it is applied on datasets such as:

1) Data collection, which involves collecting and gathering data from different sources such as the data may be stored in a database using SQL, the data can be written on a particular paper, data can also be recorded in spreadsheets and text files. These gathered data are then preserved in a particular format that is fully electronic. The electronically stored data is then ready for accurate analysis. Learning algorithms are generally developed and improved based on the accuracy of the dataset.

2) Data exploration, where the quality of the dataset is one of the prime factors. In this stage, accessibility of data depends on its condition, such as missing data, incorrect data, noisy data, inconsistent data etc. and the properties of these data are collected because the input data is used in all other stages and, ultimately, the algorithms that are learned based on the dataset are developed to predict something important about the real-life situation. Data exploration is executed with the help of manual as well as automated activities such as data visualisation or data profiling etc.

3) Data preparation, which consists of proper formatting of data. In this stage, data are formatted in a way best suited to a particular ML

model. In the case of data formatting, it is challenging to find precise anomalies and their corresponding solutions, to minimise errors, aggregate data accurately, etc. The next phase of data preparation is to improve the quality of data by removing the duplicate values or correcting the incorrect data, etc.

4) Feature extraction is an important step where feature selection is the initial step and then extraction takes place. In the case of feature selection, a group or subset of the variable on which learning algorithm is truly dependent is selected. In the case of feature extraction, it is necessary to efficiently choose those particular features or attributes or properties that are smaller in size and richer in characteristics. These extracted features are useful for purposes of further analysis.

5) Model training is executed using two types of dataset such as a training dataset and a test dataset where the training set of data are used to train the algorithms and after that the test data set is fed to the algorithm so that the algorithm makes a logic/prediction based on the given new data, which are then analyzed for accuracy. This accuracy is very important for accepting the purpose of the ML algorithm. Until and unless the accuracy is not well accepted, the training process of the algorithm will not stop.

6) Evaluation of the performance of the model where checks have been made that the model is properly and correctly working based on its applicable algorithm. The accuracy of the model is measured in this stage based on the performance of the model.

7) Improvisation of the performance of the model that has actually taken place when the model gives an inaccurate or improper result or outcome in the evaluation phase. In this phase, the learning algorithms, which are the pillar of any model, are updated again and again until it reaches its satisfactory level and shows the correct outcome (Nithya and Ilango 2017).

It is clear that performance is the primary criterion to evaluate ML algorithms that are used to build certain task-specific models, which is why a model is not accepted unless its performance has reached an optimum level with good results. The evaluation process of these algorithms depends on certain parameters. Apart from these parameters, many algorithms have their own specific parameters such as parameters defined as configuration-based variables that are generally used to learn or make estimations from data and make predictions whenever required internally of a model. Every algorithm has the ability to optimise the parameters to get more accurate results. Some of the examples of parameters are the coefficients of logistic regression or linear regression, weights, activation function, number of hidden layers of the artificial neural network, support vectors of support vector machine, split point, depth of decision tree, etc.

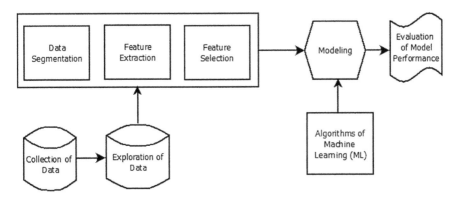

FIGURE 4.3
Different steps of ML.

Figure 4.3 shows different steps of ML whenever it is applied to a dataset. Hyper-parameters are defined as also configuration-oriented variables that are applied to a model externally and used for heuristics purpose. Some of the examples of hyper-parameters are the rate of learning for the training of a neural network, k in the k-nearest neighbor, etc. The effectiveness of the parameters of ML algorithms depends on certain metrics that are helpful for discovering the accuracy and correctness of a model.

1) Confusion Matrix: This is the easiest and most instinctive metric that is used to discover the efficiency of a model. There are certain terms that are related to confusion matrix such as True Positive (TP), where the predicted class as well as the actual class of the data point is one, True Negative (TN) where the predicted class as well as the actual class of the data point is one, False Positive (FP) where the predicted class of the data point is one and the actual class of the data point is zero and False Negative (FN) where the predicted class of the data point is zero and the actual class of the data point is one.

2) Sensitivity: It is recognised as a mensuration factor and is used to discover the exact proportion of a group who had already some particular issue and have identified those issues using specific algorithms.

$$\text{Sensitivity} = \frac{TP}{TP + FN} \tag{1.1}$$

3) Specificity: This is identified as a measurement element through which the actual proportion of a group who did not have some specific issue is also identified as not having those issues.

$$\text{Specificity} = \frac{TN}{TN + FP} \tag{1.2}$$

4) Accuracy: Defined as a total number of accurate predictions from a group of different types of predictions (Chakraborty et al. 2016).

$$\text{Accuracy} = \frac{TP + TN}{TP + TN + FP + FN} \qquad (1.3)$$

Positive Predictive Value (PPV) is the proportion of positive results in the area of diagnostics tests and statistics that are actually true positive results whereas the Negative Predictive Value (NPV) is also the proportion of negative results in those areas. Both PPV and NPV are used to describe the true performance of statistical measures or any diagnostic test.

5) False Alarm Rate: It can be generated at a particular moment when a noise or any other signal that has occurred, either from a malfunctioning portion of equipment or from the internal or external sources of radar, exceeds a predefined detected threshold level. This False Alarm rate may be defined as an audio signal, an output of a digital signal processor or a blip on a Cathode Ray Tube (CRT). If the threshold is very low then there will be a large number of false alarms and if the threshold is very high then the total numbers of false alarms are very low.

6) Receiver Operating Characteristics (ROC) Curve: This curve is used to describe the diagnostic ability of a classifier system where the detecting threshold is always varied. This ROC curve is defined to accurately plot the True Positive Rate (TPR) in different settings of threshold against False Positive Rate (FPR).

7) Area Under the ROC (AUC) Curve: This curve is used to describe an aggregate-based measure of the overall performance of all kinds of thresholds, which are defined as classification-based thresholds. The ranges of this curve are between 0 and 1. AUC is used to measure the proper quality of the prediction of a particular model without bothering precisely what type of threshold is chosen and also used to measure good predictions. The AUC value is 0.0 when the prediction of a model is 100% incorrect and the value is 1.0 when the prediction of a model is 100% correct.

4.6 Supervised Learning Methods

In the case of supervised learning, the prime objective is to predict the desired output based on the known procedure. This supervised learning is defined as a predictive model where the prediction can be done with the help of values within a dataset and an efficient algorithm. This predictive model gives a thorough idea of its requirement through a specific procedure that shows how the learning procedure takes place and what the aim is for doing this learning process; that is why training such a model which is identified as a predictive is known as supervised learning. The aim of supervised learning is to construct a prediction-based model, which is used to accurately predict the target output with the help of the evidence that exists in uncertainty. The algorithm of supervised learning is based on a known input–output data pair as a training process and then with the help of the outcome of the training process, it trains a predictive model in such a way that this model is used to generate the realistic and accurate predictions based on new data (Nithya and Ilango 2017). Supervised learning is categorised into several techniques or algorithms such as KNN, Random Forest, Decision Tree, Naive Bayes, SVM, ANN, different types of regression techniques like linear regression, logistic regression, polynomial regression, etc. The training of supervised learning algorithms is done based on labeled examples where these algorithms receive a collection of inputs with their corresponding exact outputs in a training phase and then the algorithms learn through a comparison between the actual output and received output with the help of errors based on a new data set in a test phase. Supervised learning uses the features from labeled data for the prediction of unlabeled data. This learning is mostly used in applications where historical data is used to predict the events that will happen in the future, such as the learning procedure can easily predict fraudulent card transactions or which customer has filed an insurance claim. This learning is an established and important area of ML, whose highest contribution is in the healthcare sectors. Nowadays, this learning is widely used for the diagnosis of cancer where much medical data related to cancer are collected and then used to properly train the particular system. Whenever a new specific case is assigned to the system, the system tries, at an optimum level, to predict whether the tumor is benign or malignant based on its particular size and other pre-defined factors. Figure 4.4 shows how training images are used to properly train a machine and how the proper prediction takes place using the test image.

4.7 Unsupervised Learning Methods

Unsupervised learning is applied to perform a particular task that would be beneficial in cases where no particular prediction model is used. The objective of unsupervised learning is to recognise the input data and then discover some structure or features within it. This is mostly used for transactional data. It is categorised into different techniques or algorithms such as K-Means clustering, Principle Component Analysis (PCA), Singular Value

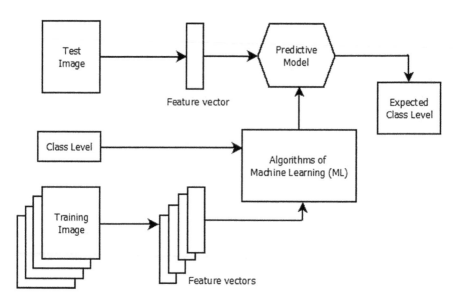

FIGURE 4.4
Supervised learning method.

Decomposition (SVD), Apriori association analysis, Hidden Markov model, etc. These algorithms are efficiently used to propose different items, identify proper data-based outliers and also segment the text topics (Nithya and Ilango 2017). This learning is used for the type of data that does not have any historical or previous labels. The algorithms of unsupervised learning are used to first explore the input data and then to discover the structure within these data. This learning method is used to recognise previously unexplored predictors.

4.8 Reinforcement Learning (RL)

RL is defined as an efficient learning process where a machine is trained to make accurate decisions using the requirements of business, along with the aim of maximising performance. RL discovers its target through a trial and error process. This learning procedure is used to save time and minimise human involvement. RL is mostly used in robotics technology like navigation or gaming sectors etc. (Nithya and Ilango 2017). It consists of three basic components such as the decision-maker or learner or agent, the overall environment to which an agent can interact and the actions that an agent can perform. This scheme enhances the desired reward using the perfect actions which are done by an agent within a limited period of time. In the case of RL, the learning procedure that is achieved with the help of the maximising ability of a defined reward is influenced by behavioral psychology. Examples in this category are the self-driving cars of Google, where reinforcement learning is used to improve the knowledge of different routes or the Lilly drone camera where the drones have self-learned to fly correctly and then to capture images using sensors. The types are RLisSarsa, Actor-Critic method, and Q-learning, etc.

4.9 Semi-Supervised Learning

The application area of semi-supervised learning is the same as that of supervised learning, and semi-supervised learning is used for both unlabeled data and labeled data. This learning procedure is useful where the labeling cost is very high to permit a training process with fully labeled dataset. Semi-supervised learning is necessary in specific cases, such as when recognising a particular person's face through a webcam (Nithya and Ilango 2017). It is defined as intermediate learning, between unsupervised and supervised learning, which takes a collection of input data, where a small group of these input data have their associated labels. In the case of semi-supervised learning, the labeled and unlabeled data are combined and used to predict the desired output and also to organise the data for better performance. This learning is actually a cost-effective procedure because a massive amount of labeled data that is used in supervised learning is very costly and it also requires less execution time. An example of semi-supervised learning is Facebook's face-recognition technique. These learnings are categorised as Graph-based methods, Generative models, etc. Most of these learning algorithms are used in the medical field in various ways but some categories of supervised learning are widely used in healthcare sectors. Supervised ML categories such as KNN, Naive Bayes (NB), DT, SVM, ANN are widely used to provide unique characteristics for accurate diagnosis, exact detection, proper treatment and recovery sectors in the healthcare field.

4.9.1 K-Nearest Neighbor (KNN)

K-Nearest Neighbor (KNN) algorithm, also defined as lazy learning, instance-based learning, case-based reasoning, example-based reasoning or memory-based reasoning algorithm, is the most commonly used algorithm used to store input cases or training data instances and then, using the classification technique, to identify new cases using similar measures. In KNN, using the nearest neighbor technique, the total number of training data that is closest to a new point on the basis of distance criteria is chosen and then the proper level from the training data instances (Jolly and Gupta 2019). KNN algorithm is basically used to select K most similar instances in the dataset which are properly classified to an input query and classifies the query into proper class using the majority class of k selected data.(Agrawal 2016). The performance of K-NN is measured using computation that depends on similar criteria (Park and Lee 2018). The value of K is the primary factor on which the measurement of KNN is done. Here K is used to identify the total number of nearest neighbor, which is assigned for the classification of a training data set. These training datasets are responsible for assigning weights according to distance from the sample dataset. KNN is divided into two groups such as structure-based NN technique and structureless NN technique. The general, commonly used KNN falls into the category of the structureless NN technique. In the case of the structure-based NN technique, different data structures are involved like a k-d tree, orthogonal structure tree (OST), axis tree, ball tree, etc. (Pradip and Deorankar 2018). In the medical field, KNN is useful in several areas but, through its classification technique, it is mainly used to accurately predict and improve the condition of heart disease, which always requires a precise diagnosis and proper treatment because it is one of the major causes of death. Heart disease consists of several categories such as coronary heart disease, cardiovascular disease, inflammatory heart disease, cardiomyopathy, Hypertensive heart disease, etc. KNN is efficiently used to provide correct results and good performance in systems designed and developed to effectively diagnose heart disease (Deekshatulu and Chandra 2013). KNN, along with genetic algorithms, serves major contribution in medical sectors such as medicine or chemical drugs. This combination is also used in cardiology, endocrinology, oncology, gynecology, radiology, obstetrics, surgery, neurology, infectious diseases and orthopedics (Pradip and Deorankar 2018).

4.9.2 Naive Bayes (NB)

NB is also one of the important categorisations of supervised learning that is widely applicable for the analysis and characterisation of any uncertainty within the real-time domains. In the case of NB, quantitative-based measurements are used to recognise uncertainty that occurs in different models. This network is used to encode the relations of the variables with the help of graph theory and probability (Reiz and Csató 2009). NB, along with its associated techniques, are generally applicable for reasoning and capturing using uncertainty. The areas where the usability of NB is beneficial are the decision-making system, PPDM, and the reliability estimation system, image processing, spam filtering, the healthcare field as well as biomedicine, etc. NB consists of unique features to handle the missing and incomplete data that are normally found in medical records, for accurate prediction using precision. NB gives a tremendous performance in the medical field for cases of accurate prediction and is therefore used for effective diagnoses such as risk analysis in the medical ground, dental pain, BPD, etc. (Agrawal 2019). Several types of survival prediction related to surgery are also achieved using NB.

In the case of clinical epidemiology, NB is used to construct models of different diseases and in the case of bioinformatics; NB is truly beneficial for the interpretation of microarray-based gene expression data. NB is also very useful to properly predict outbreaks (Jolly and Gupta 2018). This network takes time series of corresponding electronic health records to accurately track the proper patterns of epidemics. NB is also used in several real-life healthcare-based sectors such as gene regulatory networks, bio-monitoring, proper documentation and retrieval of medical information, system-based biology, medicine, accurate classification, semantic search, etc.

4.9.3 Decision Trees (DT)

DT, which is also as a prime division of supervised learning, is used to represent rules that are understandable and accepted by humans and also used in knowledge-based systems. DT is actually a tree-structure-based flowchart where each node represents a specific test on the value of an attribute, each branch is used to represent the outcome of a specified test and tree leaves are used to represent class distributions. The specific instance is categorised from the root or parent node of the tree. The objective of DT is basically to predict the particular membership of several objects that belong to different classes using the values that correspond to their predictor variables or attributes. DT consists of a group of cases that are already solved. These solved set cases are categorised into two divisions, such as a training set, which is applicable for induction purpose of DT and a test set, which is appropriate to verify the obtained solution's accuracy. In the case of clinical decision analysis, some factors are very important. First, the outcome of clinical decision analysis is not only applied to a single patient but also to a group of patients. Second, in the case of decision-making, all the reports and documents of each patient must be taken into account. Third, whenever a decision is taken, the patient's opinion is very important (Peng et al. 2018). DT has been widely used in healthcare and medical applications for many years. DT models with simplicity, conceptual capability and the possibility of automatic learning features are the appropriate models for executing any decision-related work in the medical field. DT is mainly used for diagnosis and classification of disease in healthcare sectors because in some cases, diagnosis actually requires continuous monitoring of electronic neuropathy. Many real-life situations in the medical field need efficient decisions and, in this case, DT is a very effective and reliable method that is used to provide high-classification accuracy using the knowledge which is gathered and represented as a simple format (Batra and Agrawal 2018). The proper identification or recognition of automatic cardiovascular neuropathy is the prime factor in the diagnosis of diabetic disease. In the case of an electronic record, the collected data are properly analyzed through DT. This type of analyzed data is used in treatment plans and advanced diets (Gharehchopogh et al. 2012). DT is also used to help the medical expert take critical decisions at a crucial time in an efficient manner.

4.9.4 Support Vector Machine (SVM)

SVM is another important division of supervised learning whose popularity has increased regularly through its performance. This algorithm is widely applicable for classification as well as regression tasks that are used to properly analyze and then identify patterns. SVM has been developed with the help of statistical learning theory and associated learning algorithms (Kampourakia et al. 2013). SVM is a model-free and data-driven approach that generally separates two groups or classes using proper generation of the specific hyperplane through which the two groups are divided only when the incoming data are mathematically transformed in a high-dimensional space. SVM has a great impact on healthcare sectors, especially in the biomedical and bioinformatics fields (Sweilam et al. 2010). This technique is used to design and develop methods that are used to automatically classify and detect diseases. SVM has great potential to perform and improve systems in the case of epidemiologic study, which generally includes several risk factors such as genome-based profiles based on gene expression and association data, limited knowledge of the biological relationship, limitation in sample size, etc. For various types of complex and critical diseases where the interactions between gene and environment or gene and gene are considered, SVM is an effective modeling technique that ensures proper classification to detect complex diseases such as diabetes (Yu et al. 2010). In the case of patients who are suffering from critical diseases such as heart disease or cancer, SVM has provided accurate prediction and proper detection in these cases (Son et al. 2010). The medical data that are relevant to the patient do not always contain meaningful information, and in most cases the medical data contain incomplete information, missing values, redundant information, noisy data, etc. These types of data create lots of problems in the case of preprocessing and in later steps. Analysis of medical data is very important in the case of proper diagnosis and if the data is not correct, due to noise or incompleteness, then it directly affects analysis, which is vital for the prediction of disease. SVM is used to solve these problems through its unique characteristic, which is capable of analyzing and predicting any disease using incomplete information. Medical experts are really satisfied with the helpfulness of SVM because it is really critical in diagnosing the correct disease without analysis of medical data (Weng et al. 2016).

4.9.5 Artificial Neural Network (ANN)

The interconnection of neurons in the animal brain is modeled in modern-day computing technology, termed as an ANN. The computing systems composed of ANN possess the ability to gather knowledge about performing tasks from data inputs using their AI without being specifically programmed to do so. Analogous to animal brains, ANN also consists of artificial neurons. These interconnected neurons transmit a signal to each other. Any particular neuron can receive, process and transmit the signal further to any other neurons. The objective of building ANN was to incorporate the manner of human brains for solving problems into computing systems. Recently it has been successfully applied to fields like machine translation, gaming, speech recognition, image analysis, etc. and many more real-world problems. The ANN mainly consists of three types of layers. There are mainly two operations involved through these layers of an ANN: Learning and Recall. Learning is the process of adjusting the connection weights according to some specific learning rule when some specific learning examples are applied as an input buffer. The recall is the process of accepting input and producing a particular response determined by the learning of the network. There are four main characteristics that make neural networks distinct from other traditional computing and AI technology: (i) learning by example, (ii) fault tolerance, (iii) distributed associative memory and (iv) pattern recognition. The storage of a unit of knowledge is distributed across all the connection weights in an ANN. Due to this associative property even on the application of a partial input, the trained network will discover the closest match and produce a corresponding output that is nearly full. The information in ANN is distributed in several processing elements throughout the network. So the failure of any single element will cause a negligible effect in the functioning of the whole network. The inherent superiority of ANN makes it especially efficient in recognising patterns and creating new ones more easily. Due to this ability, ANN can match a large amount of input simultaneously with a reasonable response to noise and produce a categorised or generalised output. There are six types of ANN used in ML recently: Feed-Forward Neural Network, Radial basis function Neural Network, Kohonen Self Organising Neural Network, Recurrent Neural Network (RNN), CNN, Modular Neural Network. The frequently used areas of ANN are image analysis and compression, character recognition, target classification, noise filtering, text-to-speech conversion, etc. ANN is the backbone of any modern ML algorithms.

4.10 Deep Learning (DL)

The computational models consisting of multiple processing layers use DL as a tool to learn the representation of data with different abstraction levels. Intricate structures in a huge data set are discovered by DL using backpropagation algorithms to suggest the modification of internal parameters of a machine to compute the representation of data in a certain layer from its previous layer representation. The primary purpose of DL is the proper credit assignment in a neural network for the exhibition of the desired behavior. It learns through several computational stages the accurate measurement of the credit to be assigned to the neural network. In a more abstract way, DL

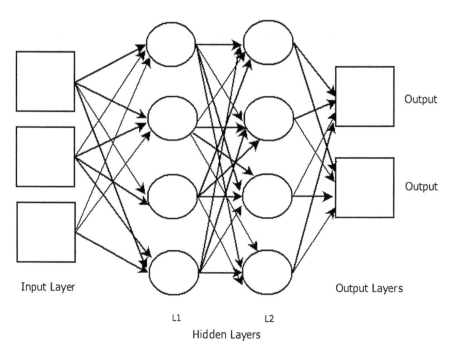

FIGURE 4.5
Deep learning applications.

is a way to allow the computer to learn with a hierarchy of concepts. The machine gathers knowledge by experiencing and understanding in terms of this concept, which is defined through its relation to the simpler ones in the hierarchy. This approach eliminates the need for any human intervention to make the machine knowledgeable for any particular computation. Speech recognition, object detection, visual object recognition, etc. are the domains where DL has recently been most successful. Figure 4.5 shows the architectural diagram of DL that is widely used in the field of healthcare.

4.10.1 Recurrent Neural Network (RNN)

AnRNN is a special type of ANN in which connections between the nodes form a directed graph along a temporal sequence. As a result, temporal dynamic behavior is exhibited. In RNNs sequence of inputs is processed using the internal state of the memory. In this type of neural network output from the previous step is fed as input to the current step. The RNN came into existence when it was needed to know previous words when predicting a word in making a sentence. This purpose of remembering previous words is solved by a hidden layer in RNN. Thus, the hidden-state feature of RNN remembers the information about sequences. RNN possesses a memory to remember this information. In the case of RNN, as it performs the same task on every input and hidden layer, each of them also uses the same parameters. So, unlike other neural networks, the complexity of parameters is reduced in RNN.

4.10.2 Convolutional Neural Network (CNN)

As we use RNN for predicting the sequence of words in a sentence, CNN is used for image classification. In CNN or convents, parameters are shared all along the network. It assigns learnable weights and biases as important to various aspects or objects in an image in order to differentiate one from another. The major advantages of convents are lower pre-processing overhead than other classification algorithms and its exclusive ability to learn the filters and characteristics eliminating the overhead of hand-engineering these filters with required training. The convents are inspired by the visual cortex organisation in the human brain and analogous to the connectivity patterns of the neurons. The eminence of convents lies in capturing the spatial and temporal dependencies in an image using relevant filters. Due to the reduced number of parameters and reusability of weights, it performs a better fit to the image dataset. In a more abstract way, convents are made to better understand the sophistication of an image more easily with proper training.

4.10.3 Deep Learning in Healthcare

The huge volume of big data generation in medical fields has created the requirement for new technology to analyze and extract information effectively from this enormous amount of data. DL is evolving rapidly to fulfill this purpose efficiently. The inherent capability of more rapid computational power makes DL exceedingly successful in interpreting and gaining insight from highly heterogeneous medical data like medical imaging, clinical genomics and electronic health records from highly dispersed sources. DL can extract features and learn patterns from this large data set quite easily. The most commonly used algorithms of deep learning in medical applications are CNN, RNN, Deep Belief Network (DBN). DL is presenting itself as a leader in medical image analysis. In recent research for identifying skin cancer from clinical images, it has proved its success with sensitivity and specificity over 90%. CNN, which is a DL algorithm, has emerged as a deep image mining to detect referable diabetic retinopathy through retinal fundus photographs. DL gives a better result in processing aggregated Electronic Health Records (EHRs) than any other conventional ML algorithms. DL is also applied to model continuous-time signals, such as laboratory results, to automatically identify specific phenotypes and predict disease onsets from laboratory test measures alone. Researchers have proposed DL as a successful tool to reconstruct the true genome sequences and call genetic variants from high-throughput sequencing data like RNA or DNA sequencing, gene expression and epigenetic studies.

4.11 Existing Works on ML for Biomedical and Health Informatics

Vellido et al. (2012) discussed the interpretation process for different ML models which is one of the important characteristics of every ML technique. Di and Joo (2007) discussed how a wireless sensor network (WSN) is influenced by ML. This paper showed a clear picture how various ML techniques are used to solve application and networking-related problems such as energy constraints, bandwidth constraints, resource allocation, etc. in WSN in detail. Singh et al. (2012) reported the different application of optical character recognition (OCR), which is used in the healthcare field to most efficiently handle lots of documents such as the health-related papers or the insurance-related papers of each individual patient. Cruz et al. (2006) surveyed different ML-based applications that are used for the correct diagnosis of cancer. In this paper, the explanation, comparison and performance between various ML techniques that are efficiently used for cancer prediction are well discussed. Libbrecht et al. (2015) presented a complete overview of different applications based on ML which are beneficial for various genomics and genetics applications such as analysis of metabolomics or proteomics data, annotations of sequence elements and epigenetic, etc. Yang et al. (2015) proposed a framework as well as a platform that is used to design cloud-based health monitoring and also manufacturing systems with the help of ML. Jiang et al. (2017) discussed the latest and upcoming status of AI in healthcare segments. This work showed how the popular and well-known AI technique known as ML is used for diagnosis and prediction of various diseases or to discuss several applications in medical fields. Feldman et al. (2017) presented an overall impact of ML in the healthcare field, which focused on several challenges such as the beginning of preprocessing followed by process modeling and then the interpretation of the desired output. The several uncountable ML-based challenges and applications in health informatics segments have been discussed (Holzinger 2016). Clifton et al. (2012) described ML-based health informatics systems that are applicable in different phases of the journey of a patient such as continuous monitoring of a patient in a general ward, ultimate care in an intensive care unit (ICU), constant tracking of the past and present records of each individual patient. Liu et al. (2016) reported several methods of extracting biomedical information, which is very useful in various applications such as integrative biology and clinical-based decision support. Ravi et al. (2017) described elaborate discussion on how one well-defined technique of ML known as DL is combined with health informatics and used in the medical field in different ways such as automatic feature generation, reduction of human intervention, different data-driven solutions. Here, the advantages and disadvantages of several deep learning methods such as an RNN, deep belief networks are well described. This work also discussed different applications of health informatics such as gene variants, diagnosis of cancer, classification of neural cells, clustering of cells, detection of the tumor, classification of brain tissues, segmentation of organs, selection and classification of the gene, which are truly influenced and benefited by DL. Holzinger and Jurisica (2014) provided an overview of integrative and interactive solutions for the discovery of knowledge along with some prime challenges, which include the necessity of developing different algorithms and models for preprocessing, fusion, mapping, interpretation, analysis of critical data in healthcare segments. Miotto et al. (2017) suggested that the well-known approach of ML known as DL is actually beneficial in multiple phases of the healthcare sector where the applications of DL methods create enormous advantages in several clinical domains and also the analysis of the massive volume of biomedical data. A computer-based screening system (DREAM), used to analyze and detect the fundus images to identify the presence of any diabetic retinopathy (DR), with the help of ML (Roychowdhury et al. 2014), has been proposed. Different types of techniques or classifiers such as SVM, Gaussian mixture model (GMM), KNN are used to classify red lesion and bright lesion or to classify retinopathy and non-retinopathy lesion. A computer-based classification framework is used to detect and differentiate different types of brain tumors combing the perfusion MRI with conventional MRI (Zacharaki et al. 2009). This proposed scheme consists of several steps wherein the feature extraction stage, and different features such as intensity or shape of the tumors are included. Do et al. (2018) proposed a system that is used to detect melanoma disease with an optimum level of accuracy using the resources available in smartphones. This work also introduced the latest numerical features that are used to identify specific skin lesion and also proposed an improved algorithm based on feature selection that is used to extract a tiny set of discriminative features. An effective system that is used to classify skin images to identify the skin lesions from the malignant and using ML is implemented (Roslin and Janney 2018). The important part in curing melanoma disease is to properly classify and detect the particular category of melanoma. In this paper, different types of classifiers are used for accurate melanoma classification such as decision tree, KNN, random forest, SVM, multilayer perceptron (MLP) with an accuracy of 86.9%, 71.5%, 93%, 69% and 75% respectively. Oliveira et al. (2018) described different classification methods of ML such as Bayesian network, SVM, decision tree, KNN, ANN, which are used to classify pigmented skin lesion. Georgakopoulos et al. (2018) introduced a new approach, already tested within two real-life applications such as diagnosis of several structures at a superpixel state and classification of dermoscopy images into non-malignant and malignant cases, with the help of a division of ML known as a CNN. Galvez et al. (2018) proposed a methodical approach using ML that consists of the integration of different heterogeneous series of skin cancer and the design of multiclass classifiers to efficiently and accurately distinguish cancer-affected skin from normal skin. Majtner et al. (2018) introduced an improved method that is used to detect melanoma where the

DL concept and the linear discriminant analysis (LDA) are jointly used to decrease the dimensionality of feature space and increase the accuracy of classification phase. Here, DL is actually used to optimise the features for accurate detection of melanoma. Peng et al. (2018) proposed a segmentation-based architecture that consists of a discrimination network and U-net (U-net is a full CNN) to efficiently execute the segmentation task because this task consists of several problematic issues such as skin contrast or hair obstruction. Brinker et al. (2018) discussed a methodical review regarding the classification phase of skin cancer using CNN. A melanoma detection method using a simple technique that is used for hair removal purposes in the preprocessing stage and also DL method is considered for classification purpose (Salido and Ruiz 2018). In this work, the improved preprocessing technique is used to provide excellent performance in terms of accuracy, sensitivity and specificity. Authors (Lee et al. 2018) designed a pipeline named as WonDerM, which is used to resample the images of skin lesion that are already preprocessed, to construct a fine-tuned architecture of neural network using segmentation-based task data and, lastly, to classify seven types of skin diseases with the help of the ensemble method. Chen et al. (2017) proposed a new system that is based on CNN (which is the prime division of DL where DL also is a subset of ML). Here, CNN is used to predict the risk of multimodal-oriented diseases with the help of unstructured and structured data and provides an accuracy of 94.8% with a minimum amount of time. Wagle et al. (2013) introduced a model based on the classification technique using an improved KNN that is used to identify the retinal fundas medical images in a much better way. An algorithm combines a genetic algorithm and KNN to efficiently classify the heart disease. Through this algorithm, effective prediction and early detection of heart disease can be performed easily and the risk factors related to any disease will also decrease (Deekshatulu and Chandra 2013). The latest system using e-health-based cloud servers, where the privacy of datasets such as a patient's diagnosis results or personal information are the prime issues, is also used for medical diagnostic purposes in an efficient way (Park and Lee 2018). The comparative study between SVM and KNN classifiers, which are widely and efficiently used to predict heart disease, has found that accuracy is much better compared to KNN (Shilpa et al. 2017). A KNN along with a genetic algorithm is used as a combined means of classification and improvisation of drugs and chemical medicines (Pradip and Deorankar 2018). Jabbar (2017) suggested an approach based on KNN and Particle swarm optimisation (PSO), which is used to predict heart disease where the classifier is applied to decrease the rate misclassification. PSO algorithm for chronic wound-tissue classification using LDA method under tele-wound networks is proposed and provides 98% accuracy (Chinmay 2017). Khamis et al. (2014) discussed the application of KNN classifier in medical or healthcare data because medical data are a very important factor for accurate diagnosis, and improper analysis of these data can cause harm or even death. Lee and Abbott (2003) presented a discussion about BN as a tool for knowledge discovery that is used by nurse researchers. Iqbal et al. (2015) surveyed different BN representations, learning as well as predictions procedures and several inferences that are drawn easily from BN representations and different types of DN applications in various domains. Reiz and Csató (2009) implemented a new classifier based on BN that has been invented to take correct decisions related to medical or healthcare sectors and also accurately analyze medical data to save human life. Gharehchopogh et al. (2012) discussed several applications of DT in the healthcare domain, such as DT, which serves an important role while making decisions in healthcare domains. Jong (2014) described several analyses based on clinical decisions using a classification technique known as DT. Clinical analysis using DT has the ability to resolve uncertainty and complexity in the medical field. Podgorelec et al. (2002) discussed different properties of DT in detail and several DT-based approaches with suitable emphasis on existing and future applications in the medical domain. Sweilam et al. (2010) surveyed a comparative study of how SVM is used for the diagnosis of cancer. A new approach is proposed using SVM that is applicable to these medical cases to properly classify persons with or without particular diseases, such as the correct classification of pre-diabetes and diabetes patients (Yu et al. 2010). An application system is considered that is web-based and known as "e-doctor" and is used to automatically predict a diagnosis using SVM in the medical domain (Kampouraki et al. 2013). Son et al. (2010) introduced an SVM-based system that is used to classify the predictors of medication adherence in the case of patients suffering from heart failure. Janardhanan et al. (2015) analyzed the usefulness and outstanding performance of SVM when it is used in medical data for accurate diagnosis purposes. This work showed that when several different types of dataset are applied to a predictive model, the analyzing accuracy of SVM is far better compared to other classifiers. Chakraborty (2017, 2019) considered ML approaches for chronic wound-tissue characterisation that provide good classification accuracy. A new learning method is considered that is incremental in nature, based on SVM and which is used in several medical data related applications (Wenget et al. 2016).

4.12 Conclusion and Future Issues

The objective of the chapter is to give an overall guide to the healthcare sector where different ML techniques provide extraordinary and unexpectedly good results. This chapter also provides a clear idea of how to explore different applications, recent trends and the impact of ML in the medical domain and future issues. In the past, when healthcare lacked the various applications of ML, it was impossible to handle large medical data along with its different abnormal features like noisy data, inappropriate data, redundant data. Nowadays with different use of ML, it is easy to handle multiple sectors in the healthcare domain, such as the feature selection method, which is very helpful in increasing the accuracy and performance of the various classifiers. Many people who are directly or indirectly related to the healthcare fields, such as patients, doctors, organisations and medical insurers, have benefited from the wide use of ML. Presently, patients are able to get treatment that is very cost-effective, medical insurers are able to identify fraud cases, doctors or medical practitioners are able to easily and quickly identify the accurate disease and also prescribe the correct medication. ML is gaining popularity and is widely used in the healthcare domain due to its inexhaustible descriptive and analytical capability, and because this technology consistently provides improved services, such as enhancement of patient and medical services, with minimum cost, and provision of fast emergency services.

There are several issues that create challenges for ML researchers, such as collaboration between different entities like ML researchers, hospital, vendors, patient, insurers. This type of collaboration is absolutely beneficial to the improvement of the standard of healthcare because this collaboration has is able to resolve the issue of data unavailability. Next, the issue is that nowadays large volume of medical data is constantly generated, which is why more efficient and sophisticated methods are needed for the medical industry to accurately and properly handle that massive volume of data. One of the major challenges is to overcome the issue of medical data annotations in cases of rare or complicated diseases. This issue will be resolved if supervised learning is moved efficiently to unsupervised or semi-supervised learning without affecting performance. An extremely important issue is privacy and security. The healthcare segment is constantly dealing with large amounts of data, most of which is patient data. These data, which are related to patient health, are confidential and cannot be shared to anyone except the medical expert. It is a big challenge for researchers to handle these confidential medical data with full privacy. The primary issue is to create efficient techniques that are extremely beneficial for prediction as well as decision-making purposes because in the healthcare field taking a good decision can save a patient's life. The challenge is to build efficient virtual assistance or continuous monitoring for elderly persons, physically challenged people and sick people to provide them with continuous healthcare assistance. Another issue is the improvisation of learning algorithms to obtain improved performance of prediction. In the near future, if ML applications are successfully developed and properly implemented it would not only be valuable for computer-based healthcare systems but also medical experts will benefit through the emerging technologies that will facilitate and improve the quality and efficiency of the healthcare sector.

References

Agrawal, R. 'A modified K-nearest neighbor algorithm using feature optimization'. *Int. J. Eng. Technol* 8, no. 1 (2016): 28–37.

Agrawal, R. 'Predictive analysis of breast cancer using machine learning techniques'. *Ingeniería Solidaria* 15, no. 3 (2019): 1–23.

Batra, M. and Agrawal, R., 2018. 'Comparative analysis of decision tree algorithms'. In *Nature inspired computing* (pp. 31-36). Springer, Singapore.

Brinker, T. J, Hekler, A., Utikal, J. S, Grabe, N., Schadendorf, D., Klode, J., Berking, C., Steeb, T., Enk, A. H. and Von, K. C, 2018, 'Skin cancer classification using convolutional neural networks: systematic review', *Journal of Medical Internet Research*, vol. 20, no. 10.

Chakraborty, C., 2017, 'Chronic wound image analysis by particle swarm optimization technique for tele-wound network', *International Journal of Wireless Personal Communications*, vol. 96, no. 3, pp. 3655–3671. doi: 10.1007/s11277-017-4281-5.

Chakraborty, C., 2019, 'Computational approach for chronic wound tissue characterization', *Elsevier: Informatics in Medicine Unlocked*, pp. 1–10. doi: 10.1016/j.imu.2019.100162.

Chakraborty, C., Gupta, B., Ghosh, S. K., Das, D. and Chakraborty, C., 2016, 'Telemedicine supported chronic wound tissue prediction using different classification approach', *Journal of Medical Systems*, vol. 40, no. 3, pp. 1–12.

Chen, H., Fuller, S. S., Friedman, C. and Hersh, W., 2005, 'Knowledge management, data mining, and text mining in medical informatics', in *Medical informatics*, Springer, Boston, MA, 3–33.

Chen, M., Hao, Y., Hwang, K., Wang, L. and Wang, L., 2017, 'Disease prediction by machine learning over big data from healthcare communities', *IEEE Access*, vol. 5, pp. 8869–8879.

Chinmay, C., 2017, 'Chronic wound image analysis by particle swarm optimization technique for tele-wound network', *International Journal of Wireless Personal Communications*, vol. 96, no. 3, pp. 3655–3671.

Clifton, D. A., Gibbons, J., Davies, J. and Tarassenko, L., 2012, 'Machine learning and software engineering in health informatics', in *Proceedings of the first international workshop on realizing AI synergies in software engineering*, IEEE Press, 37–41.

Clifton, D. A., Niehaus, K. E., Charlton, P. and Colopy, G. W., 2015, 'Health informatics via Machine learning for the clinical management of patients', *Yearbook of Medical Informatics*, vol. 24, no. 01, pp. 38–43.

Cruz, J. A. and Wishart, D. S., 2016, 'Applications of ML in cancer prediction and prognosis', *Cancer Informatics*, vol. 2, p. 117693510600200030.

Deekshatulu, B. L. and Chandra, P., 2013, 'Classification of heart disease using k-nearest neighbor and genetic algorithm', *Procedia Technology*, vol. 10, pp. 85–94.

Di, M. and Joo, E. M., 2007, 'A survey of ML in wireless sensor networks from networking and application perspectives', in *6th international conference on information, communications & signal processing*, IEEE, pp. 1–5.

Do, T. T., Hoang, T., Pomponiu, V., Zhou, Y., Chen, Z., Cheung, N. M. and Tan, S. H., 2018, 'Accessible melanoma detection using smartphones and mobile image analysis', *IEEE Transactions on Multimedia*, vol. 20, no. 10, pp. 2849–2864.

Fang, R., Pouyanfar, S., Yang, Y., Chen, S. C. and Iyengar, S. S., 2016, 'Computational health informatics in the big data age: a survey', *ACM Computing Surveys (CSUR)*, vol. 49, no. 1, p. 12.

Feldman, K., Faust, L., Wu, X., Huang, C. and Chawla, N. V., 2017, 'Beyond volume: the impact of complex healthcare data on the machine learning pipeline. Towards integrative machine learning and knowledge extraction', *LNCS*, vol. 10344, pp. 150–169.

Gálvez, J. M., Castillo, D., Herrera, L. J., San Román, B., Valenzuela, O., Ortuño, F. M. and Rojas, I., 2018, 'Multiclass classification for skin cancer profiling based on the integration of heterogeneous gene expression series', *PloS one*, vol. 13, no. 5, p. e0196836.

Georgakopoulos, S. V., Kottari, K., Delibasis, K., Plagianakos, V. P. and Maglogiannis, I., 2018, 'Improving the performance of convolution neural network for skin image classification using the response of image analysis filters', *Neural Computing and Applications*, pp. 1–18.

Gharehchopogh, F. S., Mohammadi, P. and Hakimi, P., 2012, 'Application of decision tree algorithm for data mining in healthcare operations: a case study', *International Journal of Computer Applications*, vol. 52, no. 6, pp. 21–26.

Holzinger, A., 2016, 'Interactive machine learning for health informatics: when do we need the human-in-the-loop?', *Brain Informatics*, vol. 3, no. 2, pp. 119–131.

Holzinger, A. and Jurisica, I., 2014, 'Knowledge discovery and data mining in biomedical informatics: the future is in integrative, interactive machine learning solutions', in *Interactive knowledge discovery and data mining in biomedical informatics*, Springer, pp. 1–18.

Iqbal, K., Yin, X. C., Hao, H. W., Ilyas, Q. M. and Ali, H., 2015, 'An overview of Bayesian network applications in uncertain domains', *International Journal of Computer Theory and Engineering*, vol. 7, no. 6, p. 416.

Jabbar, M. A., 2017, 'Prediction of heart disease using k-nearest neighbor and particle swarm optimization', *Biomedical Research*, vol. 28, no. 9, pp. 4154–4158.

Janardhanan, P., Heena, L. and Sabika, F., 2015, 'Effectiveness of support vector machines in medical data mining', *Journal of Communications Software and Systems*, vol. 11, no. 1, pp. 25–30.

Jiang, F., Jiang, Y., Zhi, H., Dong, Y., Li, H., Ma, S. and Wang, Y., 2017, 'Artificial intelligence in healthcare: past, present, and future', *Stroke and Vascular Neurology*, vol. 2, no. 4, pp. 230–243.

Jolly, S. and Gupta, N., 2018, 'Extemporizing the data trait', *International Journal of Engineering Trends and Technology*, vol. 58, no. 2, pp. 100–103.

Jolly, S. and Gupta, N., 2019, 'Handling mislaid/missing data to attain data trait', *International Journal of Innovative Technology and Exploring Engineering*, vol. 8, no. 12, pp. 4308–4311.

Jong, M.-B., 2014, 'The clinical decision analysis using decision tree', *Epidemiology and Health*, vol. 36, p. e2014025.

Kampourakia, A., Vassisa, D., Belsisb, P. and Skourlasa, C., 2013, 'e-Doctor: a web based support vector machine for automatic medical diagnosis', *Procedia – Social and Behavioral Sciences*, vol. 73, pp. 467–474. doi: 10.1016/j.sbspro.2013.02.078.

Khamis, H. S., Cheruiyot, K. W. and Kimani, S., 2014, 'Application of k-nearest neighbour classification in medical data mining', *International Journal of Information and Communication Technology Research*, vol. 4, no. 4, pp. 990–1000.

Lee, S. M. and Abbott, P. A., 2003, 'Bayesian networks for knowledge discovery in large datasets: basics for nurse researchers', *Journal of Biomedical Informatics*, vol. 36, no. 4–5, pp. 389–399.

Lee, Y. C., Jung, S. H., Won, H. H. and WonDer, M, 2018, 'Skin Lesion Classification with Fine-tuned Neural Networks', *arXiv preprint arXiv:1808.03426*.

Libbrecht, M. W. and Noble, W. S., 2015, 'Machine learning applications in genetics and genomics', *Nature Reviews Genetics*, vol. 16, no. 6, p. 321.

Liu, F., Chen, J., Jagannatha, A. and Yu, H., 2016, 'Learning for biomedical information extraction: methodological review of recent advances', *arXiv preprint arXiv:1606.07993*.

Majtner, T., Yildirim-Yayilgan, S. and Hardeberg, J. Y., 2018, 'Optimised deep learning features for improved melanoma detection', *Multimedia Tools and Applications*, pp. 1–21.

Miotto, R., Wang, F., Wang, S., Jiang, X. and Dudley, J. T., 2017, 'Deep learning for healthcare: review, opportunities, and challenges', *Briefings in Bioinformatics*, vol. 19, no. 6, pp. 1236–1246.

Nithya, B. and Ilango, V., 2017, 'Predictive analytics in health care using machine learning tools and techniques', in *IEEE international conference on intelligent computing and control systems (ICICCS)*, pp. 492–499.

Oliveira, R. B., Papa, J. P., Pereira, A. S. and Tavares, J. M. R., 2018, 'Computational methods for pigmented skin lesion classification in images: review and future trends', *Neural Computing and Applications*, vol. 29, no. 3, pp. 613–636.

Park, J. and Lee, D. H., 2018, 'Privacy preserving k-nearest neighbor for medical diagnosis in e-health cloud', *Journal of Healthcare Engineering*, pp. 8–11.

Peng, Y., Wang, N., Wang, Y. and Wang, M., 2018, 'Segmentation of dermoscopy image using adversarial networks', *Multimedia Tools and Applications*, pp. 1–17.

Podgorelec, V., Kokol, P., Stiglic, B. and Rozman, I., 2002, 'Decision trees: an overview and their use in medicine', *Journal of Medical Systems*, vol. 26, no. 5, pp. 445–463.

Pradip, A. S. and Deorankar, A. V., 2018, 'Classification of chemical medicine or drug using K nearest neighbor (kNN) and genetic algorithm', *International Research Journal of Engineering and Technology*, vol. 5, no. 3, pp. 833–834.

Rahul, C. D., 2015, 'Machine learning in medicine', *Circulation*, vol. 132, no. 20, pp. 1920–1930. doi: 10.1161/CIRCULATIONAHA.115.001593.

Ravì, D., Wong, C., Deligianni, F., Berthelot, M., Andreu-Perez, J., Lo, B. and Yang, G. Z., 2017, 'Deep learning for health informatics', *IEEE Journal of Biomedical and Health Informatics*, vol. 21, no. 1, pp. 4–21.

Reiz, B. and Csató, L., 2009, 'Bayesian network classifier for medical data analysis', *International Journal of Computers Communications and Control*, vol. 4, no. 1, pp. 65–72.

Roslin, E. and Janney, B., 2018, 'Classification of melanoma from dermoscopic data using machine learning techniques', *Multimedia Tools and Applications*, pp. 1–16.

Roychowdhury, S., Koozekanani, D. D. and Parhi, K. K., 2014, 'DREAM: diabetic retinopathy analysis using ML', *IEEE Journal of Biomedical and Health Informatics*, vol. 18, no. 5, pp. 1717–1728.

Salido, J. A. A. and Ruiz, Jr C., 2018, 'Using deep learning to detect melanoma in dermoscopy images', *International Journal of Machine Learning and Computing*, vol. 8, no. 1, pp. 61–68.

Shilpa, J., Abdul, A. and Vanitha, T., 2017, 'A comparative study on data classification algorithms knn and SVM in diagnosing heart disease', *International Journal of Latest Trends in Engineering and Technology*, Special Issue - SACAIM, vol. 2017, pp. 366–369.

Singh, A., Bacchuwar, K. and Bhasin, A., 2012, 'A survey of OCR applications', *International Journal of Machine Learning and Computing*, vol. 2, no. 3, p. 314.

Son, Y. J., Kim, H. G., Kim, E. H., Choi, S. and Lee, S. K., 2010, 'Application of support vector machine for prediction of medication adherence in heart failure patients', *Healthcare Informatics Research*, vol. 16, no. 4, pp. 253–259.

Sweilam, N. H., Tharwat, A. A. and Moniem, N. A., 2010, 'Support vector machine for diagnosis cancer disease: a comparative study', *Egyptian Informatics Journal*, vol. 11, no. 2, pp. 81–92.

Tom, M. M, 2006, 'The discipline of machine learning', *CMU-ML-06-108*.

Vellido, A., Martín-Guerrero, J. D. and Lisboa, P. J., 2012, 'Making machine learning models interpretable', *ESANN*, vol. 12, pp. 163–172.

Wagle, S., Mangai, J. A. and Kumar, V. S., 2013, 'An improved medical image classification model using data mining techniques', in *7th IEEE GCC conference and exhibition*, pp. 114–118.

Weng, Y., Wu, C., Jiang, Q., Guo, W. and Wang, C., 2016, 'Application of support vector machines in medical data', in *4th international conference on cloud computing and intelligence systems (CCIS)*, IEEE, pp. 200–204.

Yu, W., Liu, T., Valdez, R., Gwinn, M. and Khoury, M. J., 2010, 'Application of support vector machine modeling for prediction of common diseases: the case of diabetes and pre-diabetes', *BMC Medical Informatics and Decision Making*, vol. 10, no. 1, p. 16.

Yang, S., Bagheri, B., Kao, H. A. and Lee, J., 2015, 'A unified framework and platform for designing of cloud-based machine health monitoring and manufacturing systems', *Journal of Manufacturing Science and Engineering*, vol. 137, no. 4, p. 040914.

Zacharaki, E. I., Wang, S., Chawla, S., SooYoo, D., Wolf, R., Melhem, E. R. and Davatzikos, C., 2009, 'Classification of brain tumor type and grade using MRI texture and shape in a machine learning scheme', *Magnetic Resonance in Medicine: An Official Journal of the International Society for Magnetic Resonance in Medicine*, vol. 62, no. 6, pp. 1609–1618.

5

Meta-Heuristic Algorithms: A Concentration on the Applications in Text Mining

Iman Raeesi Vanani and Setareh Majidian

CONTENTS

5.1 Introduction

A cognitive system is a system with a corpus as a main element of body knowledge and takes various types of data to ingest them. In many cognitive systems textual data consists of a huge amount of body knowledge: the data are books, articles, comments, notes and reports. To analyse data accumulated in the corpus of the cognitive system, data analytics techniques are generated (Hurwits et al., 2015). In the field of data analytics algorithms, meta-heuristic algorithms are deployed in different areas. Clustering

(Kuo et al., 2018) (Nayak et al., 2017), classification (Alatas, 2012), sentiment analysis (Ahmad et al., 2015), Opinion mining (Saraswathi andand Tamilarasi, 2012) and feature selection (Wang et al., 2012) take advantage of meta-heuristic algorithms. Meta-heuristic algorithms which feature the use of a community increase the optimum level of achievement (Nanda and Panda, 2014). Meta-heuristic algorithms are deployed in global optimisation problems. A meta-heuristic algorithm has two main features of "intensification" and "diversification", or "exploitation" and "exploration". Diversification refers to developing diverse solutions because of searching globally. However, intensification shows the focus of search on local points by finding local solutions in the region. The combination shows the best solution (Wang et al., 2012).

Despite wide application for meta-heuristic algorithms it is found that there is a lack of a comprehensive approach towards the algorithms Researchers try to create a full view of meta-heuristic algorithms. In this research we focus on identifying meta-heuristic algorithms with the approach of text mining. Genetic Algorithms (GA), Ant Colony Optimisation (ACO), Ant lion optimiser (ALO), Bat Algorithm (BA), Cat swarm optimisation algorithm (CSO), Crow search algorithm (CSA), Cuckoo Optimisation Algorithm (COA), Bee Colony Optimisation (BCO), Particle swarm optimisation (PSO), Firefly algorithm (FA) and Tabu Search Algorithm (TS) have been introduced in this research and the application of algorithms in text mining has been identified where present. The chapter is ordered as a literature review section, which investigates meta-heuristic algorithms, then the eleven types are introduced. In Section 3 a model based on meta-heuristic has been proposed and discussed for text mining applications. In Section 4, future research is considered, and the final section is a conclusion.

5.2 Literature Review of Meta-Heuristic Algorithms

Nature-inspired algorithms divide into two groups: evolutionary and swarm intelligence. The first is related to Darwin's evolutionary theory. It proposes that phenomenon with a high level of adaptability have a better chance of surviving in comparison with those which this feature. The second considers the natural behaviour of creatures (Fister et al., 2015). They are powerful and popular sources which have become the focus of much research in recent decades. They offer a wide number of solutions to a candidate problem (Hatamlou, 2013). It is dependent on the evaluation of potential solutions, which is one of the main advantages of meta-heuristic algorithms, however, it is not required to have detailed prior knowledge of the system (Thomas and Jin, 2014). Meta-heuristic algorithms offer global or near optimal solutions, considering acceptable time and search costs (Fong et al., 2017). Meta-heuristic algorithms are effective in many fields like load-balancing problems (Milan et al., 2019) and schedule problems (Younis and Yang, 2018). Meta- heuristic algorithms have the potential for robust adaptive behavior in searching processes, which is the result of the neighbourhood function (Chavez et al., 2019). It is proven that meta-heuristic algorithms are effective in solving complex and large-scale problems (Sahoo et al., 2018). Meta-heuristic algorithm hybridisation is performed in two ways: loosely coupled and strongly coupled. The former refers to sequential methods, where the output of each level is the input of the next level and the final result is the output of the final level through saving their identity; however, the latter is about using a method as the main method and the others playing supporting roles (Younis and Yang, 2018).

Meta-heuristic algorithms can handle "discrete variables", "multi-objective optimisation problems" and "non-linear" problems. Classic optimisation methods offer a unique solution in each iteration with a local, but not global, view and with a random local starting point with predefined rules (Reddy and Bijwe, 2016).

5.2.1 Genetic Algorithms (GA)

GA has been introduced as a randomised search that aims to find near-optimal solutions to problems with a high level of complexity and dimension. The main parameter in GA that is considered in space searches is chromosomes with string shape. It is also called population, in the case of generating a collection. Random population will be created through a degree, which shows the efficacy and fitness function related to each string. The output of this step is to select a string and copies of it, which will be entered into the mating pool. Mutation processes and crossover create a new string, which has routed to the previous string. The process will be continued until a termination condition is found. According to previous research, "image processing", "neural network" and "machine learning" are just a few examples of GA (Lee and Lee, 2015). Genetic algorithms have routed in on genetic and natural selection algorithms (Jan et al., 2017). The focus of GA is to find optimal solutions and optimal clustering but not to consider starting points and clustering metrics (Lee and Lee, 2015). In the feature selection process GA takes two main approaches, namely, filter and wrapper, which focuses on taking heuristic-based data characteristics to find the value of features like correlation and evaluate the efficacy of the GA solution that is completed by a machine learning algorithm (Peng et al., 2015). However, K-means algorithms offer optimised local-point solutions by initialising seed values that are entered to cluster generation, and GA deploys to near-optimal or optimal clustering solutions based on searches done on the base of initial seed. The comparison between the two methods shows that GA outperforms K-means (Babu and Murty, 1993). GA also demands attention for creating the "classifier system" and the "mining association rules" (Ruan and Zhang, 2014). According to literature on GA, feature selection needs strong tools as the data has become overwhelmed with lots of features and finding sufficient amounts of feature is an issue in text-mining tasks. Performance has divided into two groups: dependent and independent from learning. The first is related to the wrapper approach and the latter takes the filter approach. Despite the fact that the filter approach is independent from the learning algorithm, there is a possibility of the dependency of the optimal set of features to the learning algorithm. This is considered as one of the main drawbacks of the method. However, the wrapper approach outperforms the filter approach by deploying learning algorithms in evaluation of every feature set. Complexity in computation filed is one of the main problems of the approach, which is solved by deploying GA in feature selection as a learning algorithm (Ramsingh and Bhuvaneswari, 2018).

5.2.2 Ant Colony Optimisation (ACO)

Dorigo proposed the ant colony optimisation method in 1991. ACO has been introduced as a population-based stochastic method (Ding et al., 2015). The method imitates real ant behavior in the process of searching for food. A bionic algorithm will be chosen to find the optimal solution (Mavrovounioti and Yang, 2015). A brief description of the method is that when ants start to seek food, they deposit a pheromone, and the amount of the material is in opposite relation to the distance between the food source and the deposited material. This is a guide for new ants that choose a greater amount of pheromone (Tack, 2018). The traveling salesman problem (TSP) was the first application of ACO (Jayaprakasha and KeziSelvaVijila, 2019). Routing Optimisation Algorithm (Lohrmann and Luukka, 2018), data mining (Jiang and Chen, 2016), robot path planning (Tang et al., 2018) and deep learning (Li et al., 2016) are just a few examples of ACO applications.

Creating a strong bond between the method and other algorithms, effective performance of the algorithms in optimisation, but not swarm intelligence, are some advantages of ACO. It also allows work on distributed parallel computing with high levels of speed and accuracy, with the aim of finding quasi-optimal solutions (Lohrmann and Luukka, 2018). Ants try to improve their solutions through a local search based on the emitted material (German et al., 2014). The first step of ACO is to initialise the pheromone trail, which is followed by generating the solution through building the pheromone trail. It then evolves into a trail pheromone. ACO consists of evaporation and reinforcement phases in each pheromone updating procedure. Evaporation of pheromone fractions and emitting of pheromone happens in each identified step, which is continued till termination condition has arrived (Mohan and Baskaran, 2012).

ACO has been used in textual documents as an algorithm in the feature selection process, which takes classifier performance and the length of selected feature as an input in the ACO algorithm without the need to have previous knowledge of the feature (Aghdam et al., 2009). In sentiment analysis ACO has been deployed in feature selection with the wrapper approach to solve problems of the high dimensionality of textual documents (Ahmad et al., 2019). It is also deployed in clustering documents as it is proven to have strong performance in this area (Guo et al., 2014; Dziwiński et al., 2012).

5.2.3 Ant Lion Optimiser (ALO)

ALO is a newly proposed swarm intelligence algorithm that is inspired by ant lion behaviour. It performs on the basis of a stochastic global search algorithm (Wang et al., 2019). It was first proposed by Mirjalili but some ineffectiveness of ALO, like "slow convergence rate" and "easy premature convergence", led to the development of the Improved Ant Lion optimiser to cover these problems (WuZhongqiang et al., 2017). The underlying assumption of the algorithm is that ant lions move in a circular track in sand pits, using their jaws to trap ants. They trap ants when they move into pits, then reconstruct these pits to hunt their next prey. The algorithm adapts behavior based on hunting ants (WuZhongqiang et al., 2017). ALO comprises of two main stages, namely: the larvae stage, which is responsible for hunting; and the adult stage, whose main responsibility is to reproduce. It works on the basis of two population sets of ant lions and ants. It searches for global optimal solutions in five steps:

- Random ant walking
- Trapping generation
- Trapping in ant lions' pits
- Hunting the prey
- Rebuilding the pit (Wang et al., 2018).

It deploys "power system economic load dispatching", "hydro thermal scheduling" and "optimal power flow" problems. Ants are search agents (Rajan et al., 2017).

5.2.4 Bat Algorithm (BA)

BA is inspired by the behavioral patterns of bats. Considering bat echolocation, the life span of each signal lasts just a few thousandths of a second with high-level frequency in the range of 25 to 150 kHz. One of the main rules in the algorithms is that bats can differentiate between prey and obstacles. They can find their way on the basis of the rate and frequency of sound (Tang et al., 2020). BA was proposed by Yang (2010) (Xie et al., 2019).

Fundamental BA rules are:

- Bats adapt their distance from prey through echolocation behavior
- Bats fly with velocity to the target by changing direction in a particular interval through signal imitating

- By calculating the distance to the target, bats adapt the wavelength and pulse rate
- The volume of the sound is decreasing constantly (Yildizdana and Baykan, 2020).

It is proven that the bat algorithm has the capability of facing with constrained and unconstrained problems. The effectiveness of the bat algorithm lies in the fact that it can find robust solutions in low-dimensional problems. However, its performance is decreased when faced with high-dimensional problems (Chakri et al., 2017). The prominence of the Internet of Things (IOT), which creates big data, enables analytic facilities to use knowledge hidden in the gathered data. BA with the feature of enhancing optimisation finds its way into IOT data analytics (Cui et al., 2019). Wind-speed forecasting (Nui et al., 2018), analyzing complex networks (Atay et al., 2017) and network intrusion detection (Enache and Sgârciu, 2015) are a few examples of BA applications.

A drawback of BA is its inefficiency in solving problems with high complexity and constant optimisation problems (Liu et al., 2018).

5.2.5 Cat Swarm Optimisation Algorithm (CSO)

CSO was proposed by Chu and Tsai in 2007. It imitates cat behavior. Two modes of cats behavior are considered in this algorithm: seeking and tracing. On the basis of this algorithm each cat has a specific position, which is considered in D-dimension and velocity with changing value by changing position (Fister et al., 2015). It is categorised as a stochastic and population-based theory. The seeking process consists of five operators:

- "Seeking Memory Pool (SMP)": It comprises of cat memory in relation to seeking behaviour.
- "Seeking Range of Selected Dimension (SRD)": It deploys for mutation activity considering dimensions.
- "Counts of Dimension to Change (CDC)": It relates to the number of transformations by changing dimensions.
- "Self Position Consideration (SPC)": It is defined as a Boolean value, which sustains the position and solution if the value is true.
- "Mixture Ratio (MR)": It is a small value, which shows that cats are mostly in seeking positions (Mohapatra, 2016).

It has been proven that CSO works much more effectively compared with PSO in finding global optimal solutions (Yang et al., 2015). It also has a high level of performance in solving linear and non-linear optimisation problems (Pappula and Ghosh, 2014). It can be used in classification and feature selection (Gill and Buyya, 2019). One limitation of CSO algorithm is considering two aspect of behavior which neglects of covering high level of coverage in global solution (Pappula and Ghosh, 2018).

5.2.6 Crow Search Algorithm (CSA)

The crow is a species of bird that has high performance in remembering faces and reacting to danger as they have bigger brains in comparison with other birds. They steal other birds' food and try to find the best hiding place by changing their position constantly. The most obvious drawback of the algorithm is poor behaviour in converging "local minima entrapment" (Shekhawat and Saxena, 2019).

Two main parameters contributing to CSA are the "flight length" and "awareness probability". It also suffers from a low level of adjustability (Mohammadi and Abdi, 2018). However, CSA performance in creating balance in exploration–exploitation is strong (Upadhyay and Chhabra, 2019). CSA proposed by Askarzadeh (2016), who defined the algorithm as "crow", which seeks for food in an environment. Each food source can be a potential solution. The seeker tries to find global optimal solutions, which are the best solutions. Two drawbacks of the algorithm are "premature convergence" and "weak diversity", which may occur in the face of constrained optimisation (Rizk-Allah et al., 2018).

5.2.7 Cuckoo Optimisation Algorithm (COA)

COA was proposed and mimics the breeding behavior of the cuckoo, which puts its eggs in the nests of other birds. "Simplicity", "high level of convergence" and "few numbers of parameters" made the algorithm appropriate to deploy (Yang et al., 2019). The underlying assumption for COA is that each cuckoo has a number of eggs (between 5 and 20), which are put in other birds' nests. The probability of each cuckoo egg is dependent on the level of cuckoo eggs' similarity to other eggs. Consequently, the population of a place follows of the number of cuckoos living in that place as cuckoos migrate to the goal place where eggs are located. "Digital communication signals", "university course time-tabling problem" (Tavana et al., 2018), "satellite image segmentation", "satellite signals recognition", "timetabling" and "predicting permeability" (Komaki et al., 2017), "multivariable controller design", "advanced machining processes" (Mellal and Williams, 2015) are just a few examples of the deployment of COA. If the foreign birds become aware of the cuckoo eggs, the cuckoo feels that the nest is unsafe and it seeks another nests (optimal solution) (Zhang et al., 2019). On the base of previous research, some benefits of deploying COA are the high level of speed and accuracy, accompanying a local search beside a global search and gaining solutions with high dimension and stability in convergence (Amiri and Mahmoudi, 2016).

5.2.8 Bee Colony Optimisation (BCO)

BCO has been inspired by the natural behavior of the honey bee. "Travelling Salesman Problem", "Internet Hosting Centre" and "Vehicle routing" are examples of BCO in optimisation problems. The method was proposed by Karaboga in 2005. The prominent feature of the method is simplicity, as well as being easy and the need to have few factors to control in optimisation problems. This method is most highly used in "face recognition", "high dimensional gene expression" and "speech segments classification". Three types of bees contribute to the ABC algorithm, namely, "Employed Bees (EBees)", "Onlooker Bees (OBees)" and "scout bees deployed". In the process, Ebees, which are equal to the food source, obtain the food source and pass nectar information to Obees, which are equal to the number of EBees. The information is extracted from the food source until finishing amount has been calculated. Then Scouts in exhausted food source initiate their work to find a new food source. The nectar amount is a quantitative measure which shows solution quality (Qawaqneh et al., 2017), (Gonzalez-Abril et al., 2017). Two steps of "step forward" and "step back" are comprised of the algorithm with the former finding new information and food sources for bees and the latter propagating information as new alternatives for beehives. The starting point of the process is when a bee leaves the hive to discover its full path of travel and it comes across the movement patterns of other bees, which consist of random dances of their "preferred paths". It is a foraging process previously examined by other bees who guide them to their final destination.

Moving from one node to another node will proceed until the final destination is reached (Caraveo et al., 2016). The shortest distance has the potential of selection and in the process two values of alpha (exploitation) and beta (exploration) is considered (Cano, 2017).

BCO has been identified as an effective method in combinatorial optimisation problems with high levels of complexity. So, it can be used as an evolutionary algorithm in a POS tagger (Alhasan and Al-Taani, 2018). BCO with the cited feature has the potential to be deployed in complex problems to create better solution. Cloning and fairness are two concepts generated by BCO, which gives strength to the algorithm to find a better solution (Forsati et al., 2015).

5.2.9 Particle Swarm Optimisation (PSO)

PSO has been introduced by Kennedy and Eberhart in 1995 as a stochastic population-based algorithm which is based on the natural ability of biological organisms, such as a group of animal that try together to find a desired location in a specific area. The algorithm aims to find global optimal solutions with the feature of easy deployment. The method is implemented taking few parameters. The underlying searching algorithm of the method is a strong process, which gives it a significant performance

(Sekaran et al., 2019). Solving a non-linear optimisation problem in a real-value search space is the first step of the algorithm, which uses an iterative searching procedure to find the destination as the optimal point. Each particle has a multi-dimensional searching space, which is updated by the searching experience of particles in neighbors. The optimal point will stay in the memory in each iteration. The local best solution is called pbest and the global optimal solution is called gbest (Gou et al., 2019). Each potential solution, known as a particle, has features of current position and velocity. Inertia weight is a measure to create balance between global and local search, which creates a critical feature for PSO in trading off between global and local searches in iteration (Sekaran et al., 2019). "Artificial neural network", "pattern classification" and "fuzzy control" are a few grounds for deploying PSO (Bonyadi.M.R, Michalewicz, 2017).

PSO has been deployed in feature selection processes with the aim of optimising the process, known as the best method for text classification processes and maintaining dimensions in a stable situation (Lu et al., 2015). PSO has also been deployed in text summarisation processes (Dalal and Malik, 2017), text clustering (Song et al., 2015) and feature selection (Abualigah et al., 2018).

5.2.10 Firefly Algorithm (FA)

FA was inspired by the global and blinking behavior of the firefly and is known as a meta-heuristic algorithm, proposed by Yang 2008 (Gou et al., 2019). Flashing light helps the firefly to find its mate, follow the potential goal and also protect itself from its predator. One of the features of the method is that it can work effectively on both local and global problems, finding local and global solutions respectively. It also has strong results as the number of nodes increases as it can find local optimal points simultaneously in each local community (Jain and Katarya, 2019). The algorithm is used for complex engineering problems. The brightness of flashing light emitted from fireflies depends on the distance between two fireflies and the atmospheric absorption coefficient (Dash et al., 2020). "Economic dispatch", "structural optimisation" and "image compression" are just a few examples of FA application. It also works better in comparison with PSO and GA (Gao et al., 2015). It also works in "unit commitment", "energy conservation" and "complex networks" fields (Shanthamallu et al., 2017). Fundamental rules of FA are:

- Each firefly in a community has the potential to attract all the other fireflies in that community.
- There is a negative but strong relationship between distance and attractiveness for fireflies. The brightest firefly initiates the searching process.
- The objective function of the firefly determines its brightness level (Le et al., 2019).

The limitation of FA is its weakness in diversity and efficiency of searching, as it limits itself to diagonal-based searches rather than region-based searches, which reduces the search area. It searches for the brightest firefly in the neighborhood, neglecting their distinctiveness. K-means focuses on finding limitation with consideration of diverse movement (Xie et al., 2019). FA is generalised from PSO, differential evolution (DE) and simulated annealing (SA) algorithms by adding some value to the cited methods (Liu et al., 2018). Four main issues of FA are "light intensity", "attractiveness", "distance" and "movement". In optimisation problems each firefly signifies a potential solution and the brightness shows its quality. In a firefly communication, two fireflies are compared to each other in consideration of their brightness, where the less bright one moves toward the brighter one. With the movement, the position of each firefly is changed and the process will be continued until the comparison between all the fireflies ends (Khalifehzadeh and Fakhrzad, 2019). FA based on feature selection processes work in Arabic text classification (Larabi et al., 2018), (Mohammad et al., 2016).

5.2.11 Tabu Search Algorithm (TS)

TS was proposed as a meta-heuristic algorithm (Kiziloz and Dokeroglu, 2018). In this algorithm, a flexible memory has been used to save the information gained about the potential solution, which is the core search of the method. On the basis of historical solutions and the current solution, one search direction will be determined. Memory serves an important role in protecting the solving algorithm from becoming trapped in the cycle. The algorithm has outstanding performance in "multi-product", "multi-objective" and "multistage" supply chain problems (Mohammed and Duffuaa, 2020). The best solution will be found by moving from one point to another. When TS starts the algorithm a tabu list will be generated to consider explored points. The next move is to an unexplored point, which may be worse or better than the current solution. The process will be continued until the termination condition emerges (Khanduzi and Sangaiahb, 2019). It is proven that TS works effectively as the number of space variables increase. Robustness and efficient time usage are two features of the method. TS has a tendency to find local optimal solution. TS needs to clarify seven variables:

- Setting a criterion for selection of independent variables.
- Defining a neighborhood point
- Initiating a solution
- Generating a tabu list and defining its size
- Defining admissible subsets
- Searching parameters
- Stopping condition (Hage et al., 2019; Gaast et al., 2014).

In each iteration the tabu list will be updated in such a way that the last record is deleted and a new record is added to the top of the list. To overcome the risk of ignoring solutions that will be faced by restriction defined in the tabu list, "aspiration criteria" will be considered. According to the criteria, if a point has better objective value in comparison with the best current objective value, the tabu solution will be chosen and will be removed from tabu list (Alharkan et al., 2019). TS is deployed in clustering problems in cloud computing problems (Lu et al., 2018). It also works to decrease production and inventory costs (Boutarfa et al., 2016).

5.3 Proposed Model for Application of Meta-Heuristic in Text Mining

Text mining has been identified as a new technology with the aim of extracting information from unstructured documents, which is an expanded view of the data mining approach. The importance of text mining has emerged where the main source of data is being introduced as text, such as notes, reports, emails and so on (He et al., 2013). Text mining offers natural language processing techniques in order to analysis textual documents. It also deploys machines to translate human language into machine language to understand and generate text (Eskici and Koçak, 2018). One of the subfields of text mining is opinion mining, which relates to understanding the positive inclination or negative tendency of a target group's opinion (Manochandar and Punniyamoorthy, 2018). Text classification and text clustering, which are known as vital tasks in text mining, are also used in opinion mining. The method that makes classification tasks much easier is feature selection as an optimisation problem (Lin et al., 2016; Tutkan et al., 2016). Meta-heuristic algorithms deploy in feature selection; for example, ACO, with the aim of creating approximate optimisation, is used in feature selection (Jayaprakasha and KeziSelvaVijila, 2019; Singh and Singh, 2019). Meta-heuristic algorithms have the potential to be used in clustering optimisation problems also (Zhu, 2019). According to the literature review, some meta-heuristic algorithms have been widely used in feature selection, classification, clustering processes in text mining, opinion mining (Li and Qing-Feng, 2018), sentiment analysis and comment summarisation (Mosa et al., 2017), with most focus on algorithms such as ACO, GA and BCO. However, based on other meta-heuristic algorithms identified in this research, others can be used. The application of meta-heuristic algorithms in text mining is shown in Figure 5.1. It is shown that taking one meta-heuristic algorithm, or a mixture of them, in feature selection and then using the result in clustering and classification processes helps to generate high-quality text mining and

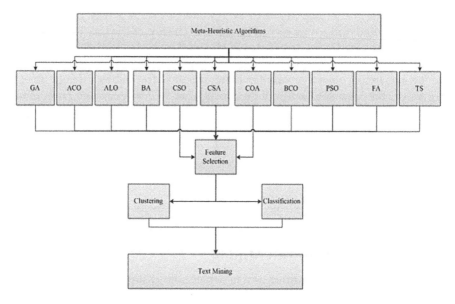

FIGURE 5.1
Meta-heuristic algorithms in the classification and clustering phase.

analysis. Meta-heuristic algorithms are those that create solutions based on two groups, namely: evolutionary and swarm intelligence and on a global scale as shown in Figure 5.1.

5.4 Future Research

According to numerous research in the area of text mining, from trend analysis to opinion mining and everything in between, a wide span of application of meta-heuristic algorithms can be identified. Extracting depressive symptoms (Wu et al., 2020), identifying tourist envioronment (Wang et al., 2020), predicting online consumer behavior (Vanhala et al., 2020), creating recommender system based on opinion mining (Da'u et al., 2020) is just some of recent research in the area of text analytics. Based on Figure 5.2, the process of text analytics initiates with gathering textual documents, which are entered into the preprocessing phase so as to ignore unwanted issues in the data. Then, based on the aim of the research, clustering, classification, or both, will be deployed to achieve the goal of the research, which can be sentiment analysis, opinion mining, trend analysis, text summarisation, topic modeling and pattern recognition. In this phase, meta-heuristic algorithms help the researcher to effectively classify, cluster, sentiment analyze or even recognise human behavior.

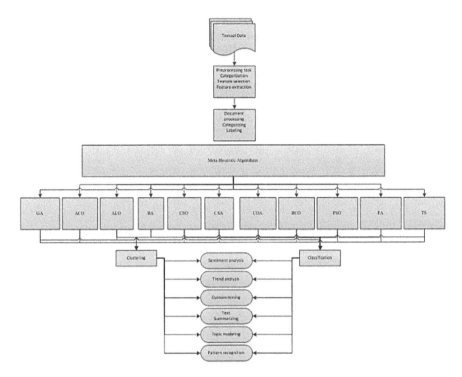

FIGURE 5.2
Future applications of meta-heuristic algorithms in text analytics. (Adapted from Packiam and Prakash, 2015.)

5.5 Conclusion

Most data usage in businesses and applications is based on text and comprises a corpus of cognitive systems from textual data; it is therefore vital for organisations to have the means to analyze text. There are many meta-heuristic algorithms that help researchers to achieve high-quality text mining. Meta-heuristic algorithms have the potential to create globally optimised solutions. However, classical methods are used for finding solutions locally. According to previous research, only a limited number of those algorithms have been used in text analytics. The contribution of this research is to introduce meta-heuristic algorithms and also to motivate researchers to deploy them in text mining, especially in clustering and classification through feature extraction. Most meta-heuristic algorithms have found application in other fields of research, such as engineering, but there is an absence of these algorithms in the area of text analytics. Meta-heuristic algorithms with the feature of finding optimal and global solutions and the potential to work on complex problems work effectively in comparison with other classic methods, which take a local approach to optimisation problems. Meta-heuristic algorithms also perform in clustering and classification problems to extract features from target textual data. In this research 11 types of meta-heurist algorithms have been identified: Ant colony optimisation, Genetic algorithms, Ant lion optimiser, Bat algorithm, Cat swarm optimisation algorithm, Crow search algorithm, Cuckoo optimisation algorithm, Bee colony optimisation, Particle swarm optimisation, Firefly algorithm and Tabu search algorithm. A light comparison between methods has been done to emphasise the important features of each method.

References

Abualigah, L. M., et al., 2018, 'A new feature selection method to improve the document clustering using particle swarm optimization algorithm', *Journal of Computational Science*, vol. 25, pp. 456–466.

Aghdam, M. H., et al., 2009, 'Text feature selection using ant colony optimization', *Expert Systems with Applications*, vol. 36, no. 3, Part 2, pp. 6843–6853.

Ahmad, S., et al., 2015, 'Metaheuristic algorithms for feature selection in sentiment analysis', *Science and Information Conference (SAI)*. doi: 10.1109/SAI.2015.7237148.

Ahmad, S. R., et al., 2019, 'Ant colony optimization for text feature selection in sentiment analysis', *Intelligent Data Analysis*, vol. 23, no.1, pp. 133–158.

Alatas, B., 2012, 'A novel chemistry based metaheuristic optimization method for mining of classification rules', *Expert Systems with Applications*, vol. 39, no. 12, 15, pp. 11080–11088.

Alharkan, I., et al., 2019, 'Tabu search and particle swarm optimization algorithms for two identical parallel machines scheduling problem with a single server', *Journal of King Saud University - Engineering Sciences*.

Alhasan, A. and Al-Taani, A., 2018, 'POS tagging for Arabic text using bee colony algorithm', *Procedia Computer Science*, vol. 142, pp. 158–165.

Amiri, A. and Mahmoudi, S., 2016, 'Efficient protocol for data clustering by fuzzy cuckoo optimization algorithm', *Applied Soft Computing*, vol. 41, pp. 15–21.

Atay, Y., et al., 2017, 'Community detection from biological and social networks: a comparative analysis of metaheuristic algorithms', *Applied Soft Computing*, vol. 50, pp. 194–211.

Babu, G. P. and Murty, M., 1993, 'A near-optimal initial seed value selection in K-means algorithm using a genetic algorithm', *Pattern Recognition Letters*, vol. 14, no. 10, pp. 763–769.

Bonyadi, M. R. and Michalewicz, Z., 2017, 'Particle swarm optimization for single objective continuous space problems: a review', *Evolutionary Computation*, vol. 25, no. 1, pp. 1–54.

Boutarfa, Y., et al., 2016, 'A Tabu search heuristic for an integrated production-distribution problem with clustered retailers', *IFAC-PapersOnLine*, vol. 49, no. 12, pp. 1514–1519.

Cano, A., 2017, 'A survey on graphic processing unit computing for large-scale data mining', *WIREs Data Mining Knowl Discover*, p. 1232. doi: 10.1002/widm.1232.

Caraveo, C., et al., 2016, 'Optimization of fuzzy controller design using a new bee colony algorithm with fuzzy dynamic parameter adaptation', *Applied Soft Computing*, vol. 43, pp. 131–142.

Chakri, A., et al., 2017, 'New directional bat algorithm for continuous optimization problems', *Expert Systems with Applications*, vol. 69, pp. 159–175.

Chavez, J., et al., 2019, 'A hybrid optimization framework for the non-convex economic dispatch problem via meta-heuristic algorithms', *Electric Power Systems Research*, vol. 177, p. 105999.

Cui, Z., et al., 2019, 'Optimal LEACH protocol with modified bat algorithm for big data sensing systems in Internet of Things', *Journal of Parallel and Distributed Computing*, vol. 132, pp. 217–229.

Dalal, V. and Malik, L., 2017, 'Semantic graph based automatic text summarization for Hindi documents using particle swarm optimization', *Information and Communication Technology for Intelligent Systems (ICTIS 2017)*, vol. 2, pp. 284–289.

Dash, J., et al., 2020, 'Improved firefly algorithm based optimal design of special signal blocking IIR filters', *Measurement*, vol. 149, p. 106986.

Da'u, A., et al., 2020, 'Weighted aspect-based opinion mining using deep learning for recommender system', *Expert Systems with Applications*, vol. 140, p. 112871.

Ding, S., et al., 2015, 'Extreme learning machine: algorithm, theory and applications', *Artificial Intelligent Review*, vol. 44, no. 1, pp. 103–115.

Dziwiński, P., et al., 2012, 'Fully controllable Ant colony system for text data clustering', in *International symposium on swarm intelligence and differential evolution*, pp. 199–205.

Enache, A. and Sgârciu, V., 2015, 'Hadoop based parallel binary bat algorithm for network intrusion detection', *International Journal of Parallel Programming*, vol. 45, no. 5, pp. 41–51.

Eskici, H. and Koçak, N., 2018, 'A text mining application on monthly price developments reports', *Central Bank Review*, vol. 18, no. 2, pp. 51–60.

Fister, I., et al., 2015, 'Planning the sports training sessions with the bat algorithm', *Neurocomputing*, Vol. 149, Part B, pp. 993–1002.

Fong, S., et al., 2017, 'How meta-heuristic algorithms contribute to deep learning in the hype of big data analytics', *Progress in intelligent computing techniques: theory, practice, and applications*, Springer Link, pp. 3–25.

Forsati, R., et al., 2015, 'An improved bee colony optimization algorithm with an application to document clustering', *Neurocomputing*, vol. 159, no. 2, pp. 9–26.

Gaast, J., et al., 2014, 'A Tabu search algorithm for application placement in computer clustering', *Computers and Operations Research*, vol. 50, pp. 38–46.

Gao, M., et al., 2015, 'Firefly algorithm (FA) based particle filter method for visual tracking', *Optik*, vol. 126, no. 18, pp. 1705–1711.

German, F., et al., 2014, 'Do retailers benefit from deploying customer analytics?', *Journal of Retailing*, vol. 90, no. 4, pp. 587–593.

Gill, S. and Buyya, R., 2019, 'Chapter 1 - Bio-inspired algorithms for big data analytics: a survey, taxonomy, and open challenges', *Big Data Analytics for Intelligent Healthcare Management*, pp. 1–17.

Gonzalez-Abril, L., et al., 2017, 'Handling binary classification problems with a priority class by using Support Vector Machines', *Applied Soft Computing*, vol. 61, pp. 661–669.

Gou, J., et al., 2019, 'A generalized mean distance-based k-nearest neighbor classifier', *Expert Systems with Applications*, vol. 115, pp. 356–372.

Guo, Y., et al., 2014, 'An Ant colony-based text clustering system with cognitive situation dimensions', *International Journal of Computational Intelligence*, vol. 8, no. 1, pp. 138–157.

Hage, R., et al., 2019, 'Optimized Tabu search estimation of wear characteristics and cutting forces in compact core drilling of basalt rock using PCD tool inserts', *Computers and Industrial Engineering*, vol. 136, pp. 477–493.

Hatamlou, A., 2013, 'Black hole: a new heuristic optimization approach for data clustering', *Information Sciences*, vol. 222, pp. 175–184.

He, W., et al., 2013, 'Social media competitive analysis and text mining: a case study in the pizza industry', *International Journal of Information Management*, vol. 33, no. 3, pp. 464–472.

Hurwitz, J., et al., 2015, *Cognitive computing and big data analytics*, Wiley, Indianapolis, Indiana.

Jain, L. and Katarya, R., 2019, 'Discover opinion leader in online social network using firefly algorithm', *Expert Systems with Applications*, vol. 122, pp. 1–15.

Jan, B., et al., 2017, 'Deep learning in big data analytics: a comparative study', *Computers and Electrical*, pp. 1–13.

Jayaprakasha, A. and KeziSelvaVijila, C., 2019, 'Feature selection using Ant Colony Optimization (ACO) and Road Sign Detection and Recognition (RSDR) system', *Cognitive Systems Research*, vol. 58, pp. 123–133.

Jiang, P. and Chen, J., 2016, 'Displacement prediction of landslide based on generalized regression neural networks with K-fold cross-validation', *Neurocomputing*, vol. 198, no. 3, pp. 40–47.

Khalifehzadeh, S. and Fakhrzad, M. B., 2019, 'A modified firefly algorithm for optimizing a multi stage supply chain network with stochastic demand and fuzzy production capacity', *Computers and Industrial Engineering*, vol. 133, pp. 42–56.

Khanduzi, R. and Sangaiahb, A., 2019, 'Tabu search based on exact approach for protecting hubs against jamming attacks', *Computers and Electrical Engineering*, vol. 79, p. 106459.

Kiziloz, H. and Dokeroglu, T., 2018, 'A robust and cooperative parallel Tabu search algorithm for the maximum vertex weight clique problem', *Computers and Industrial Engineering*, vol. 118, pp. 54–66.

Komaki, G. M., et al., 2017, 'Improved discrete cuckoo optimization algorithm for the three-stage assembly flowshop scheduling problem', *Computers and Industrial Engineering*, vol. 105, pp. 158–173.

Kuo, R., et al., 2018, 'A hybrid metaheuristic and kernel intuitionistic fuzzy c-means algorithm for cluster analysis', *Applied Soft Computing*, vol. 67, pp. 299–308.

Larabi, S., et al., 2018, 'Firefly algorithm based feature selection for Arabic text classification', *Journal of King Saud University – Computer and Information Sciences*, vol. 32, no. 3, pp. 320–328.

Le, D., et al., 2019, 'A novel hybrid method combining electromagnetism-like mechanism and firefly algorithms for constrained design optimization of discrete truss structures', *Computers and Structures*, vol. 212, pp. 20–42.

Lee, I. and Lee, K., 2015, 'The Internet of Things (IoT): applications, investments, and challenges for enterprises', *Business Horizons*, vol. 58, no. 4, pp. 1–10.

Li, J., et al., 2016, 'Medical big data analysis in hospital information system', *Big Data on Real-World Applications*.

Li, Y. and Qing-Feng, l., 2018, 'Opinion mining for multiple types of emotion-embedded products/services through evolutionary strategy', *Expert Systems with Applications*, vol. 99, pp. 44–55.

Lin, K., et al., 2016, 'Feature selection based on an improved cat swarm optimization algorithm for big data classification', *The Journal of Supercomputing*, vol. 72, no. 8, pp. 3210–3221.

Liu, Q., et al., 2018, 'A novel hybrid bat algorithm for solving continuous optimization problems', *Applied Soft Computing Journal*, vol. 73, pp. 67–82.

Lohrmann, C. and Luukka, P., 2018, 'A novel similarity classifier with multiple ideal vectors based on k-means clustering', *Decision Support Systems*, vol. 111, pp. 27–37.

Lu, Y., et al., 2015, 'Improved particle swarm optimization algorithm and its application in text feature selection', *Applied Soft Computing*, vol. 35, pp. 629–636.

Lu, Y., et al., 2018, 'A Tabu search based clustering algorithm and its parallel implementation on Spark', *Applied Soft Computing*, vol. 63, pp. 97–109.

Manochandar, S. and Punniyamoorthy, M., 2018, 'Scaling feature selection method for enhancing the classificationperformance of Support Vector Machines in text mining', *Computers and Industrial Engineering*, vol. 124, pp. 139–156.

Mavrovounioti, M. and Yang, S., 2015, 'Training neural networks with ant colony optimization algorithms for pattern classification', *Journal of Soft Computing*, vol. 19, no. 6, pp. 1511–1522.

Mellal, M. and Williams, E., 2015, 'Cuckoo optimization algorithm with penalty function for combined heat and power economic dispatch problem', *Energy*, vol. 93, Part 2, pp. 1711–1718.

Milan, S., et al., 2019, 'Nature inspired meta-heuristic algorithms for solving the load-balancing problem in cloud environments', *Computers and Operations Research*, vol. 110, pp. 159–187.

Mohammad, A., et al., 2016, 'Integrated bisect K-means and Firefly Algorithm for hierarchical text clustreing', *Journal of Engineering and Applied Science*, vol. 11, no. 3, pp. 522–527.

Mohammed, A. and Duffuaa, S., 2020, 'A tabu search based algorithm for the optimal design of multi-objective multi-product supply chain networks', *Expert Systems with Applications*, vol. 140, p. 112808.

Mohammadi, F. and Abdi, H., 2018, 'Applied soft computing: a modified crow search algorithm (MCSA) for solving economic load dispatch problem', *Applied Soft Computing*, vol. 71, pp. 51–65.

Mohan, B. and Baskaran, R., 2012, 'A survey: Ant Colony Optimization based recent research and implementation on several engineering domain', *Expert Systems with Application*, vol. 39, pp. 4618–4627.

Mohapatra, P., 2016, 'Microarray medical data classification using kernel ridge regression and modified cat swarm optimization based gene selection system', *Swarm and Evolutionary Computation*, vol. 28, pp. 144–160.

Mosa, A., et al., 2017, 'Ant colony heuristic for user-contributed comments summarization', *Knowledge-Based Systems*, vol. 118, pp. 105–114.

Nanda, S. and Panda, G., 2014, 'A survey on nature inspired metaheuristic algorithmsfor partitional clustering', *Swarm and Evolutionary Computation*, vol. 16, pp. 1–18.

Nayak, J., et al., 2017, 'Hybrid chemical reaction based metaheuristic with fuzzy c-means algorithm for optimal cluster analysis', *Expert Systems with Applications*, vol. 79, no. 15, pp. 282–295.

Nui, T., et al., 2018, 'Multi-step-ahead wind speed forecasting based on optimal feature selection and a modified bat algorithm with the cognition strategy', *Renewable Energy*, vol. 118, pp. 213–229.

Packiam, P. and Prakash, V., 2015, 'An empirical study on text analytics in big data', in *IEEE international conference on computational intelligence and computing research (ICCIC)*.

Pappula, L. and Ghosh, D., 2018, 'Cat swarm optimization with normal mutation for fast convergence of multimodal functions', *Applied Soft Computing*, vol. 66, pp. 473–491.

Pappula, P. and Ghosh, A., 2014, 'Linear antenna array synthesis using cat swarm optimization', *AEU - International Journal of Electronics and Communications*, vol. 68, no. 6, pp. 540–549.

Peng, H., et al., 2015, 'An unsupervised learning algorithm for membrane computing', *Information Sciences*, vol. 304, pp. 80–91.

Qawaqneh, Z., et al., 2017, 'Age and gender classification from speech and face images by jointly fine-tuned deep neural networks', *Expert Systems with Applications*, pp. 78–86.

Rajan, A., et al., 2017, 'Weighted elitism based Ant Lion Optimizer to solve optimum VAr planning problem', *Applied Soft Computing*, vol. 55, pp. 352–370.

Ramsingh, J. and Bhuvaneswari, V., 2018, 'An efficient map reduce-based hybrid NBC-TFIDF algorithm to mine the public sentiment on diabetes mellitus – a big data approach', *Journal of King Saud University – Computer and Information Sciences*.

Reddy, S. and Bijwe, P., 2016, 'Efficiency improvements in meta-heuristic algorithms to solve the optimal power flow problem', *International Journal of Electrical Power and Energy Systems*, vol. 82, pp. 288–302.

Rizk-Allah, R., et al., 2018, 'Chaotic crow search algorithm for fractional optimization problems', *Applied Soft Computing*, vol. 71, pp. 1161–1175.

Ruan, X. and Zhang, Y., 2014, 'Blind sequence estimation of MPSK signals using dynamically driven recurrent neural networks', *Neurocomputing*, vol. 129, pp. 421–427.

Sahoo, K., et al., 2018, 'On the placement of controllers in software-Defined-WAN using meta-heuristic approach', *Journal of Systems and Software*, vol. 145, pp. 180–194.

Saraswathi, K. and Tamilarasi, A., 2012, 'A modified metaheuristic algorithm for opinion mining', *International Journal of Computer Applications*, vol. 58, no. 11, pp. 43–47.

Sekaran, K., et al., 2019, 'Deep learning convolutional neural network (CNN) with Gaussian mixture model for predicting pancreatic cancer', *Multimedia Tools and Applications*, pp. 1–15.

Shanthamallu, U., et al., 2017, 'A brief survey of machine learning methods and their sensor and IoT applications', in *8th international conference on information, intelligence, systems and applications (IISA)*. doi: 10.1109/IISA.2017.8316459.

Shekhawat, S. and Saxena, A., 2019, 'Development and applications of an intelligent crow search algorithm based on opposition based learning', *ISA Transactions*, vol. 99, pp. 210–230.

Singh, U. and Singh, S., 2019, 'A new optimal feature selection scheme for classification of power quality disturbances based on ant colony framework', *Applied Soft Computing*, vol. 74, pp. 216–225.

Song, W., et al., 2015, 'A hybrid evolutionary computation approach with its application for optimizing text document clustering', *Expert Systems with Applications*, vol. 42, no. 5, pp. 2517–2524.

Tack, C., 2018, 'Artificial intelligence and machine learning | applications in musculoskeletal physiotherapy', *Musculoskeletal Science and Practice*, vol. 39, pp. 164–169.

Tang, H., et al., 2020, 'A multirobot target searching method based on bat algorithm in unknown environments', *Expert Systems with Applications*, vol. 141, p. 112945.

Tang, L., et al., 2018, 'A novel perspective on multiclass classification: Regular simplexsupport vector machine', *Information Sciences*, 480.

Tavana, M., et al., 2018, 'A discrete cuckoo optimization algorithm for consolidation in cloud computing', *Computers and Industrial Engineering*, vol. 115, pp. 495–511.

Thomas, S. and Jin, Y., 2014, 'Reconstructing biological gene regulatory networks: where optimization meets big data', *Evolutionary Intelligence*, vol. 7, no. 1, pp. 29–47.

Tutkan, M., et al., 2016, 'Helmholtz principle based supervised and unsupervised feature selection methods for text mining', *Information Processing and Management*, vol. 52, no. 5, pp. 885–910.

Upadhyay, P. and Chhabra, J., 2019, 'Kapur's entropy based optimal multilevel image segmentation using Crow Search Algorithm', *Applied Soft Computing Journal*, p. 105522.

Vanhala, M., et al., 2020, 'The usage of large data sets in online consumer behaviour: a bibliometric and computational text-mining–driven analysis of previous research', *Journal of Business Research*, vol. 106, pp. 46–59.

Wang, F., et al., 2020, 'What can the news tell us about the environmental performance of tourist areas? A text mining approach to China's National 5A Tourist Areas', *Sustainable Cities and Society*, vol. 52, p. 101818.

Wang, J., et al., 2012, 'Rough set and scatter search metaheuristic based feature selection for credit scoring', *Expert Systems with Applications*, vol. 39, no. 6, pp. 6123–6128.

Wang, J., et al., 2012, 'Shape and size optimization of truss structures considering dynamic constraints through modern metaheuristic algorithms', *Expert Systems with Applications*, vol. 39, no. 20, pp. 6123–6128.

Wang, J., et al., 2018, 'An improved grey model optimized by multi-objective ant lion optimization algorithm for annual electricity consumption forecasting', *Applied Soft Computing*, vol. 72, pp. 321–337.

Wang, M., et al., 2019, 'A feature selection approach for hyperspectral image based on modified ant lion optimizer', *Knowledge-Based Systems*, vol. 168, pp. 39–48.

Wu, C.-S., et al., 2020, 'Using text mining to extract depressive symptoms and to validate the diagnosis of major depressive disorder from electronic health records', *Journal of Affective Disorders*, vol. 260, pp. 617–623.

WuZhongqiang, P., et al., 2017, 'Parameter identification of photovoltaic cell model based on improved ant lion optimizer', *Energy Conversion and Management*, vol. 151, no. 1, pp. 107–115.

Xie, H., et al., 2019, 'Improving K-means clustering with enhanced Firefly Algorithms', *Applied Soft Computing*, vol. 84, p. 105763.

Xie, X., et al., 2019, 'A novel test-cost-sensitive attribute reduction approach using the binary bat algorithm', *Knowledge-Based Systems*, p. 104938.

Yang, F., et al., 2015, 'Non-rigid multi-modal medical image registration by combining L-BFGS-B with cat swarm optimization', *Information Sciences*, vol. 316, no. 20, pp. 440–456.

Yang, Y., et al., 2019, 'Cooperative media control parameter optimization of the integrated mixing and paving machine based on the fuzzy cuckoo search algorithm', *Journal of Visual Communication and Image Representation*, vol. 63, p. 102591.

Yildizdana, G. and Baykan, O., 2020, 'A novel modified bat algorithm hybridizing by differential evolution algorithm', *Expert Systems with Applications*, vol. 141, no. 1, p. 112949.

Younis, M. and Yang, S., 2018, 'Hybrid meta-heuristic algorithms for independent job scheduling in grid computing', *Applied Soft Computing*, vol. 72, pp. 498–517.

Zhang, Z., et al., 2019, 'A hybrid optimization algorithm based on cuckoo search and differential evolution for solving constrained engineering problems', *Engineering Applications of Artificial Intelligence*, vol. 85, pp. 254–268.

Zhu, L., 2019, 'Safety detection algorithm in sensor network based on Ant colony optimisation with improved multiple clustering algorithms', *Safety Science*, vol. 118, pp. 96–102.

6

Optimizing Text Data in Deep Learning: An Experimental Approach

Ochin Sharma and Neha Batra

CONTENTS

6.1 Introduction

Deep learning follows the perception of Artificial Neural Network. Using the concept of deep learning, the models can be trained more accurately to solve different problems.

The structure of a deep learning model is based upon the structure of the brain. The brain senses input in the form of audio, text, image, video. This input is then processed and finally an output is generated. This output is nothing other than a calculated value that triggers some action to take place or is capable of providing some meaningful information.

For example, if a system has a large dataset related to textual data, this data can be categorised based on the particular entertainment, sports, news, science, cultural or spiritual information. A deep learning system can help categorise this data according to the category with the highest accuracy. Another example is in image classification, when a scanner or camera reads the data in the form of images, and information from these images can be obtained and gives the system the precise information required. A face recognition system, after capturing images from a camera, can identify a particular face from the crowd. In the same context of image classification, a system using deep learning can recognise handwritten information and digitalise it from one language to other languages as required. Having information in different languages certainly helps a large audience access content.

The most important part of deep learning is the way in which information is processed. This is divided into a number of hidden layers. Each hidden layer contains hundreds or thousands of neurons, which interact with each other and forward the processing partial output on to the next hidden layer.

During processing, these hidden layers can be arranged simply and in many different ways to obtain optimal results. So, the connection between neurons can lead to different neural network architectures. A few deep learning neural network architectures are discussed here:

1. Dense Network: In this, each neuron of a hidden layer is connected with all of the neurons of the next layer. This simply means that each neuron in one layer provides input to all the neurons of the next layer. This structure looks complex but is unbiased and experimentally achieved very good results.

2. Convolutional Neural Network (CNN): This network is used for complex classification problems like image classification. CNN reads and organises data into three dimensions – width, height and depth. Additionally, in CNN the neurons of one layer are not connected to all the neurons of the next layer but only with the selected neurons selected by the specifications during the design of the CNN model.

3. Long Short-Term Memory (LSTM): This is a recurrent neuron network model that helps to eliminate the vanishing gradient problem. This neural network model keeps all the historical data records with it. So, this model seems extremely useful in predicting time-series-related analysis. The application includes automatically predicting and writing the next text generation, generating new sequences of music based on the past learning of the ongoing music sequences etc.

If machine learning and deep learning is compared, then deep learning is based on the concept of artificial neural network. According to this, a model learns every time, even when the same number of neurons and layers are used and the same optimiser or activation function is used. On the other hand, in machine learning, the same accuracy will be observed every time for a particular algorithm. Hence, using deep learning, higher accuracy can generally be achieved. The relationship of machine learning, deep learning and artificial intelligence is shown in Figure 6.1.

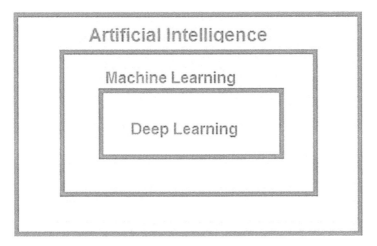

FIGURE 6.1
Relation of deep learning, machine learning and artificial intelligence.

6.2 Existing Structure of Deep Learning

Deep learning consents to train an artificial intelligence model to predict outputs, given a set of inputs. Supervised and unsupervised learning can both be thought of as advancing the artificial intelligence field. The basic work flow of deep learning is shown in Figure 6.2.

6.2.1 Neural Networks

Let's have a look in the brain using artificial intelligence.

Like humans, our estimator AI's brain has neurons. They are represented by circles. These neurons are interconnected as shown in Figure 6.3.

There are three different layers:

1. Input layer
2. Middle hidden layers
3. Final output layer

The input layer accepts initial input data. In our case, as in the diagram shown in Figure 6.3, there are neurons in the input layer. Further, the inputs are passed to the hidden layers one after another. These hidden layers do statistical computations on the received inputs. The more layers the model

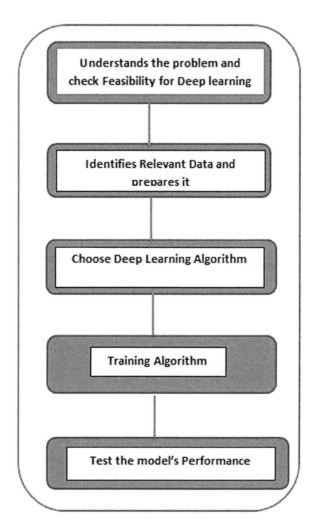

FIGURE 6.2
Types of network and basic work flow of the deep learning model.

considers, the deeper the model is. The final output layer provides the final output, generally the prediction of the data or problem related to the pre-dicted class as shown in Figure 6.4.

Every linking among neurons is connected by a weight as shown in Figure 6.5. This weight highly impacts on and determines the rank of the input value. Initially, the weights are set randomly. Further, every neuron is associated with an activation function. These functions are based on some statistical formulations. There are many types of activation functions. Choosing the correct one has a significant impact on the accuracy of the model. When all the input data has passed through all the layers of the neu-ral network, the final output is generated.

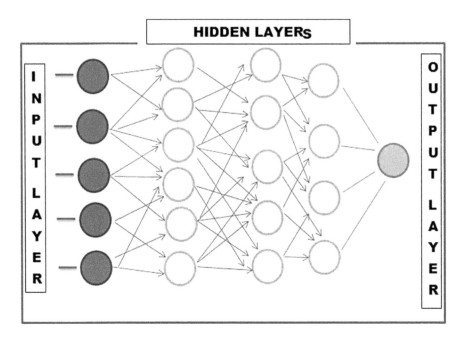

FIGURE 6.3
Neural network.

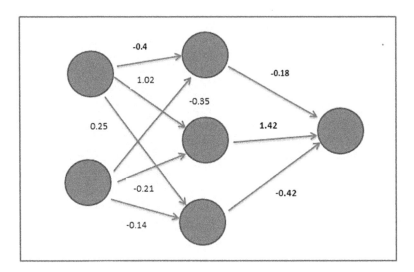

FIGURE 6.4
Neural network with weighted edges.

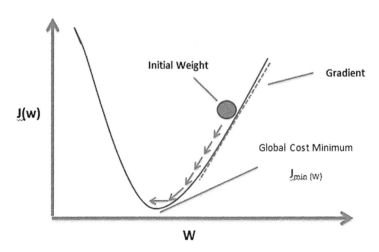

FIGURE 6.5
The role of the optimiser in deep learning.

6.3 Problems in Existing Definition

Training the AI is the hardest part of Deep Learning. The challenges can be listed as:

1. A sufficiently large data set is required.
2. A large amount of computational power is required.
3. Since the construction cost is quite high, the reduction of the cost function is set as a priority for the Deep Learning Models. For changing the cost function, we have to change the weights amid the neurons. We can change the weights randomly until the cost function is low, but this is not an effective method. In its place, we will implement a technique known as Gradient Descent. It is a technique that permits us to have the minimum function. In this case, we will look for the minimum of cost function.

   During the process we will change the weights in slight increments after every data set repetition. After calculating the gradient of the cost function at a particular set of weights, we will be able to see in which direction the minimum is.

4. To set the cost function at minimum level, we need to repeat the data set many a times. This is where the large amount of computational power is required. The updating of weights is done automatically

using the Gradient Descent. This is how Deep Learning is implemented magically. Once we've trained our airplane ticket price estimator AI, we can use it to predict future prices.

5. While developing a deep learning model, a number of inputs are to be considered, and finding non-contributing columns and dropping the non-contributing attributes are highly required.

6. Number of hidden layers: The deep learning model requires two hidden layers. One output layer is required to group the outputs from all other hidden layers. As we increase the number of hidden layers, we will, on the one hand, get the deeper analytical model and on the other hand each added layer will contribute to the computational complexity. Further, the more neurons are added to each layer the more will be the computational cost.

7. Gradient Descent Optimisers: Gradient Descent Optimisers help to minimise the cost of a model. During the designing of a model, the neural network updates its weights to get the optimum results. The optimiser function decides the change in weights based on the various factors like type and amount of data and the attributes of which optimiser function is to be used.

- Stocastic Gradient Descent (SGD): To obtain a satisfactory arrangement of weights amid input and output layers, the randomise function is used by the stochastic optimiser and for every epoch different results may be produced by the optimiser. On the basis of these results, what is optimised can be calculated.

- Adagrad, too, is a popular optimiser. Adagrad upgrades the weights manually, where the weights of attributes with similar values, or higher values or different values, are updated with the lower values.

- Adam optimiser is also known for showing good results for human language processing and computer vision. Adam has beautifully designed structure. While in SGD, each attribute and the entire neural network, the descent rate and the learning weights are the same. Adam offers diverse learning rates for various parameters and improves the overall execution of the model.

- Adadelta is an extension of Adagrad. It works similarly to Adagrad. The difference between the two is that it does not rely upon all previous gradient values, but takes a fixed size of previous gradients, which is known as a "window size" for that model.

8. Loss Function

While designing a deep learning model, a loss function helps to find the error between the actual and expected results.

$$F(\text{loss}) = \text{Expected Output} - \text{Actual Output}$$

Based on the various criteria, there are many different ways to calculate the loss function. Hence, there are various loss functions to do this, like Mean Squared Error, Mean Absolute Error, Hinge loss and Cross Entropy. Choosing an appropriate loss function in deep learning is a challenging task.

9. Activation Function

This is a function that decides whether a node's value should be considered for the final output value or if it is to be discarded. There are, statistically, many functions that can be applied as activation functions, such as sigmoid x (for binary classification), softmax, Tanh, relu, selu, softplus and swish. It has been observed that sigmoid function can be used at output activation function and shows best results with binary classification problem. Relu function will generate more dead neurons. Dead neurons cannot contribute in decision-making. But there are no hard and fast rules defined for all the activation functions and, based on their monotonicity, vanishing gradient, boundless and other functional properties and types of dataset, different accuracy can be achieved.

10. Epochs

One epoch means that the whole dataset has been run once. If, for an experiment, the epoch value is 12, this indicates that for the chosen variables, 12 times the whole dataset would be simulated. Each time, the neural network learns from it and applies this learning while simulating the next epoch. Once the optimal or consistent level of accuracy is achieved, the experiment can be suspended.

6.4 Research Trust

The following research thrusts are related to deep learning:

- Computer vision and pattern recognition
- Translation
- Create new images
- Reading text in the Wild
- Computer games, robots and self-driving cars
- Beating people in dozens of computer games
- Self-driving cars
- Robotics

- Voice generation
- Music composition
- Restoring sound in videos
- Automatically writing Wikipedia articles, math papers, computer code and even Shakespeare
- Predicting demographics and election results
- AI invents and hacks its own crypto to avoid eavesdropping
- Deep learning networks creating deep learning networks
- Predicting earthquakes

6.5 Text Classification

Text classification is the task of deciding whether a piece of text belongs to any of a set of prespecified categories. It is a generic text processing task useful in indexing documents for later retrieval, as a stage in natural language processing systems, for content analysis and in many other roles. The use of standard, widely distributed test collections has been a considerable aid in the development of algorithms for the related task of text retrieval (finding documents that satisfy a particular user's information need, usually expressed in a textual request). Text retrieval test collections have allowed the comparison of algorithms developed by a variety of researchers around the world. Standard test collections have been lacking, however, for text categorisation. Few data sets have been used by more than one researcher, making results hard to compare. The Reuters-22173 test collection has been used in a number of published studies since it was made available, and we believe that the Reuters-21578 collection will be even more valuable.

6.5.1 Steps of Text Classification

a) Problem definition: To identify the purpose and type of classification.
b) Input data
c) Creating word to index: Each word to be identified by an index.
d) Choosing max words: Out of a total dataset, choose the number of words to be processed to avoid performance issues.
e) Tokenisation: In NLP, this is an important step while pre-processing the text data. By tokenising the system, separate each word, symbol and meaningful subtle part in a string and that word can be acted on as a separate token to be matched.

f) Creating the model: Create the model by choosing the activation function, optimiser and loss function.

g) Analyze the results: Evaluate the accuracy of the designed model.

6.5.2 Developing a GUI-Based Deep Learning Application to Perform Text Classification on Reuters Dataset

In this exercise, a GUI-based tool is developed. This tool provides the facility to a user to choose from different activation functions, optimisers, loss functions and facilitate the entering of a number of neurons and neural layers while designing a deep learning model.

a) Problem definition and solution approach.

b) Input data

c) Creating word to index

```
word_index=reuters.get_word_index()#total words by index
```

d) Choosing max words

```
max_words=10000
```

e) Tokenisation

In NLP, this is an important step while pre-processing the text data. By tokenising the system separate each word, symbol and meaningful subtle part in a string and that word can be acted as separate token to get matched.

From keras_preprocessing.text**import**Tokeniser

```
tokeniser=Tokeniser(num_words= max_words)
```

f) Creating the Model

```
model.add(Activation(combo_ activation_function_for_output_laye
r.get())) ## last layer
model.compile(loss=combo_ loss_function.get(), optimiser=combo_
optimiser.get(), metrics=['accuracy'])
print("model.metrics_names",model.metrics_names)
batch_size=int(txt_value4.get())#batch_size=32
epochs=int(txt_value3.get())
```

g) Choosing the activation function, optimiser, loss function and developing the model.

h) Analyze the results.

```
score=model.evaluate(x_test,y_test,batch_size=batch_size,verbose=1)
```

6.6 Experimentation

The code and packages needed to accomplish this exercise:

- Python 3.5.2
- Pycharm editor

The software packages used in Python are:

- Keras 2.1.6
- Pandas 0.23.4
- Pip 19.0.3
- Numpy 1.16.2
- Tensorflow 1.12.3
- Scipy 1.13.0

6.6.1 Code

```
from tkinterimport* # To support GUI based approach
from tkinterimportttk
from PIL import ImageTk, Image
import keras
from keras.utils.generic_utilsimportget_custom_objects
from keras.modelsimportSequential
from keras.layersimportDense, Dropout, Activation
from keras.utils.generic_utilsimportget_custom_objects
from keras.layers.advanced_activationsimportLeakyReLU,
PReLU
import pandas as pd
from kerasimportbackend as K
from keras.datasetsimportreuters#news with 46 different
topics
window = Tk()
image1 =Image.open('C:\\Users\\ochin\\OneDrive\\Desk
top\\DOWNLOADS\m1.gif')# icon image for deep learning
optimiser window
window.title("DEEP LEARNING OPTIMISER")
window.geometry('1000x1000')
window.config(bg='yellow')
lbl_description = Label(window, text="Welcome to the
project:: Deep_LearningOptimiser.\n\n This tool
facilitates chosing among various deep learning
```

```
functionalities to run the \n\n customised experiments .
\n\n# you can choose:\t\t\t\t\t\t\t\t\t\t\t \n\n # The
desired data sets \n\n # The desired number of neural
layers\n\n # Desired number of neurons for each layer
\n\n# The desired optimiser.\n\n # The desired
activation function And can find the best results for
your problem\n", justify=LEFT,wraplength=1000,
relief=FLAT,bg= "yellow",font='Helvetica 11 bold')
lbl_description.grid(column=0, row=0)
lbl_procees = Label(window, text="\n\n\n\n DO YOU WANT TO
PROCEED ", bg="yellow",font='Helvetica 15 bold', fg="red")
lbl_procees.grid(column=0, row=1)
#########################################################
defclicked():
    btn1.destroy()
    btn2.destroy()
lbl.destroy()
lbl_procees.destroy()
window.config(bg='light blue')
#window = Tk()
window.title("Build & Test Deep_Learning Model")
window.geometry('850x600')
    box_value1 = StringVar()
combo_ activation_function_for_hidden_layer = ttk.
Combobox(window, textvariable= box_value1)
combo_ activation_function_for_hidden_layer['values']=
('softmax','selu','swish','relu','softmax','elu','ochin')
combo_ activation_function_for_hidden_layer.current(3)
combo_ activation_function_for_hidden_layer.grid(col
umn=4, row=1)
lbl_hidden_layer_act_function=ttk.Label(window,
text="Select HIDDEN layer Activation Function ")
lbl_hidden_layer_act_function.grid(column=2, row=1)
 box_value2 = StringVar()
combo_ activation_function_for_output_layer = ttk.
Combobox(window, textvariable= box_value2)
combo_ activation_function_for_output_layer['values']=
('sigmoid','softmax')
combo_ activation_function_for_output_layer.current(1)
#set the selected item
combo_ activation_function_for_output_layer.grid(col
umn=4, row=2)
lbl_output_activation_function=ttk.Label(window,
text="Select OUTPUT layer Activation Function ")
#working
lbl_output_activation_function.grid(column=2, row=2)
```

```
lbl_optimizer=ttk.Label(window, text="Select Optimizer")
#working
lbl_optimizer.grid(column=2, row=3)
   box_value22 = StringVar()
combo_ optimizer = ttk.Combobox(window, textvariable=
box_value22)
combo_ optimizer['values']= ("sgd","adam", "RMSprop",
"Adagrad", "Adadelta", "Adamax")
combo_ optimizer.current(1) #set the selected item
combo_ optimizer.grid(column=4, row=3)
lbl_loss_function=ttk.Label(window, text="Select Loss
Function") #working
lbl_loss_function.grid(column=2, row=4)
   box_value222 = StringVar()
combo_ loss_function = ttk.Combobox(window,
textvariable= box_value222)
combo_ loss_function['values']= ("mean_squared_error
","mean_absolute_error", "categorical_crossentropy",
"Adagrad", "binary_crossentropy", "mean_squared_
logarithmic_error", "mean_absolute_percentage_error")
combo_ loss_function.current(2) #set the selected item
combo_ loss_function.grid(column=4, row=4)
lbl_no_of_neurons=ttk.Label(window, text="Enter the
number of NEURONS") #working
lbl_no_of_neurons.grid(column=2, row=6)
txt_value = StringVar(value='512')
txt = Entry(window,width=10,textvariable = txt_value)
txt.grid(column=4, row=6)
lbl_no_of_hidden_layers = Label(window, text="Enter
Number of Hidden Layers" )
lbl_no_of_hidden_layers.grid(column=2, row=5)
   txt_value2 = StringVar(value='1')
txt_number_of_neurons = Entry(window,width=10,textvaria
ble = txt_value2)
txt_number_of_neurons.grid(column=4, row=5)
lbl_epochs = Label(window, text="Epochs" )
lbl_epochs.grid(column=2, row=8)
   txt_value3 = StringVar(value='2')
txt_epochs = Entry(window, width=10,
textvariable=txt_value3)
txt_epochs.grid(column=4, row=8)
lbl_batch_size = Label(window, text="Batch Size" )
lbl_batch_size.grid(column=2, row=9)
   txt_value4 = StringVar(value='32')
txt_batch_size = Entry(window, width=10,
textvariable=txt_value4)
```

```
txt_batch_size.grid(column=4, row=9)
defclick_me():
        (x_train, y_train), (x_test, y_test)= reuters.lo
ad_data(num_words=None,test_split=0.2)
word_index=reuters.get_word_index()#total words by index
print("word_index", len(word_index))#Important to
analyze dataset
print("No of training sample:{}".format(len(x_
train)))#total words for training
print("No of testing sample:{}".format(len(x_test)))
num_classes = max(y_train)+1
print("No of Classes:{}".format(num_classes))
print("x_train.shape",x_train.shape)#gives 8982 lines
print("x_train[0]", x_train[0])#training has 8982 lines,
train[0] gives first line
print("y_train.shape",y_train.shape)
print("y_train[0]", y_train[0])
print("ytrain classes for 1st row:{}".format(y_
train[0])) #this way i cud show all data in untruncated
form
print("#######################")
print("x_train[0]", x_train[0]) # print frequency wise
words indexes
print("y_train[0]", y_train[0]) # print train set class
label = 3 for 1st line of training data
        #In data set no of lines and class label is equal
and most frequent word index is class label
print("word_index['money']", word_index['money'])
index_to_word={}
for key,valueinword_index.items():
index_to_word[value]=key
print(' '.join([index_to_word[x] for x in x_train[3]]))
print(([index_to_word[x] for x in x_train[236]]))#236
line of training data
print("@@@@@")
#print(y_train.get_index_word)
max_words=10000 # out of total 21578 from dataset
from keras_preprocessing.textimportTokenizer
    tokenizer=Tokenizer(num_words= max_words)
x_train= tokenizer.sequences_to_matrix(x_train,mode=
'binary')# if word is frequent set it 1 otherwise 0
print(x_train)
x_test= tokenizer.sequences_to_matrix(x_test,mode='binary')
y_train=keras.utils.to_categorical(y_train,num_classes)
y_test=keras.utils.to_categorical(y_test,num_classes)
print(x_train.shape)
```

```
print(x_train[0])
print(y_train.shape)
print(y_train[0])
defselu1(x):
return((x)/ K.sin(x))
get_custom_objects().update({'selu1': selu1})
defswish(x):
return ((x) *K.sigmoid(x))
get_custom_objects().update({'ochin': ochin})
###################################################
model=Sequential()
i=0
p=int(txt_value2.get())
while (i< p):
i=i+1
model.add(Dense(int(txt_value.get()), input_shape=(max_
words, ), kernel_initializer='normal', activation =
combo_ activation_function_for_hidden_layer.get() ))
#model.add(Dense(64, input_shape=(max_words, ), kernel_
initializer='normal', activation= 'softplus' ))
        #model.add(Activation('ochin'))
model.add(Dropout(0.5))
model.add(Dense(num_classes)) # prediction
model.add(Activation(combo_ activation_function_for_o
utput_layer.get())) ## last layer
model.compile(loss=combo_ loss_function.get(),
optimizer=combo_ optimizer.get(), metrics=['accuracy'])
print("model.metrics_names", model.metrics_names)
batch_size=int(txt_value4.get())#batch_size=32
epochs=int(txt_value3.get())#epochs=2
history= model.fit(x_train,y_train,batch_size=batch_si
ze,epochs=epochs,verbose=1,validation_split=0.1)
        score= model.evaluate(x_test,y_test,batch_size=
batch_size,verbose=1)
#lbl4 = Label(window, text=score[0] )
lbl4 = Label(window, text=' Test Accuracy: {} '.format(s
core[1]*100),height=4).place(x=250,y=400)
#lbl4.grid(column=1, row=7, rowspan=3)
print('loss:{}'.format(score[0]))
print('Test Accuracy:{}'.format(score[1]))
action = Button(window, text="BUILD & TEST MODEL",
command= click_me).place(x=300,y=300) #working
###################################################
defclicked1():
exit()
exit()
```

```
btn1 = Button(window, text="\n\nYES", bg="yellow",
borderwidth=0, command=clicked,font='Helvetica 15
bold',fg="red")
btn1.grid(column=0, row=10)
btn2 = Button(window, text="\nNO", bg="yellow",
borderwidth=0, command=clicked1,font='Helvetica 15
bold',fg="red")
btn2.grid(column=0, row=12)
window.mainloop()
```

Figure 6.5 shows the role of the optimiser in deep learning, Figure 6.6 shows the development of GUI-based applications using deep learning and Figure 6.7 shows the accuracy prediction with a deep learning GUI-based application.

6.7 Conclusion and Future Scope

Deep learning can be a choice for various types of users to get the maximum result based on the input dataset. This field can be utilised by

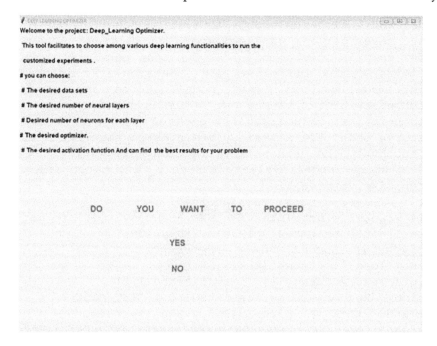

FIGURE 6.6
A deep learning GUI-based application.

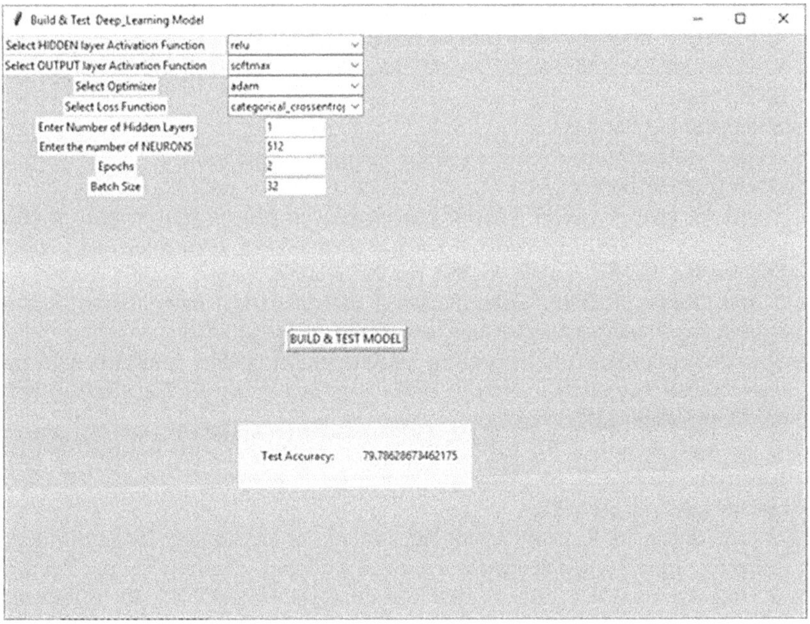

FIGURE 6.7
Accuracy prediction with a deep learning GUI-based application.

commercial online marketers and product developers where system learning is required. There are various challenges in the field of deep learning so this field is open for researchers to work on different aspects. The experiments of deep learning can be easily performed using Python and the results may be ready to pursue further the implementation and support of real-time problems.

References

Agrawal, R. and Batra, M., 2013, 'A detailed study on text mining techniques', *International Journal of Soft Computing and Engineering*, vol. 2, no. 6, pp. 118–121.

Cybenko, G., 1989, 'Approximation by superpositions of a sigmoidal function', *Mathematics of Control, Signals, and Systems (MCSS)*, vol. 2, no. 4, pp. 303–331.

Deng, L. and Yu, D., 2014, 'Deep learning: methods and applications', *Foundations and Trends® in Signal Processing*, vol. 7, no. 3–4, pp. 197–387.

Glorot, X. and Bengio, Y., 2010, 'Understanding the difficulty of training deep feedforward neural networks', in *Proceedings of the thirteenth international conference on artificial intelligence and statistics*, pp. 249–256.

Hinton, G. E., 2005, 'What kind of graphical model is the brain?', in *Proceedings of the 19th international joint conference on artificial intelligence*, pp. 1765–1775.

Hochreiter, S., Bengio, Y., Frasconi, P. and Schmidhuber, J., 2001, 'Gradient flow in recurrent nets: the difficulty of learning long-term dependencies', in S. C. Kremer and J. F. Kolen (eds.), *A field guide to dynamical recurrent neural networks*, IEEE Press.

https://keras.io/activations/.

https://www.datacamp.com/courses/deep-learning-in-python.

https://www.tensorflow.org/.

Jolly, S. and Agrawal, R., 2019, 'A broad coverage of corpus for understanding translation divergences', *International Journal of Innovative Technology and Exploring Engineering (IJITEE)*, vol. 8, no. 8S2, pp. 2278–3075.

Jolly, S. and Gupta, N., 2018, 'Extemporizing the data trait', *International Journal of Engineering Trends and Technology*, vol. 58, no. 2, pp. 100–103.

Jolly, S. and Gupta, N., 2019, 'Handling mislaid/missing data to attain data trait', *International Journal of Innovative Technology and Exploring Engineering*, vol. 8, no. 12, pp. 4308–4311.

Jolly, S. K. and Agrawal, R., 2019, 'Anatomizing lexicon with natural language Tokenizer Toolkit 3', in *Extracting knowledge from opinion mining*, IGI Global, Hershey, PA, pp. 232–266.

Kalra, V. and Agrawal, R., 2019, 'Challenges of text analytics in opinion mining', in *Extracting knowledge from opinion mining*, IGI Global, Hershey, PA, pp. 268–282.

Kaur, S. and Agrawal, R., 2018, 'A detailed analysis of core NLP for information extraction', *International Journal of Machine Learning and Networked Collaborative Engineering*, vol. 1, no. 01, pp. 33–47.

Ketkar, N., 2017, 'Introduction to tensorflow', in *Deep learning with Python*, Apress, Berkeley, CA, pp. 159–194.

LeCun, Y., Bengio, Y. and Hinton, G., 2015, 'Deep learning', *Nature*, vol. 521, no. 7553, p. 436.

McCabe, T. J., 1976, 'A complexity measure', *IEEE Transactions on Software Engineering*, vol. 2, no. 4, pp. 308–320.

Mhaskar, H. N. and Micchelli, C. A., 1994, 'How to choose an activation function', in *Advances in neural information processing systems*, pp. 319–326.

Ramachandran, P., Zoph, B. and Le, Q. V., 2017, 'Swish: a self-gated activation function', *arXiv preprint arXiv:1710.05941*, 7.

Schmidhuber, J., 2015, 'Deep learning in neural networks: an overview', *Neural Networks*, vol. 61, pp. 85–117.

Sharma, A., 2018, November, 'Understanding activation functions in neural networks', Retrieved from https://medium.com/the-theory-of-everything/understanding-activation-functions-in-neural-networks-9491262884e0.

Sharma, O., 2019, 'A new activation function for deep neural network', in *2019 international conference on machine learning, big data, cloud and parallel computing (COMITCon)*, IEEE, pp. 84–86.

Sharma, O., 2019, 'Deep challenges associated with deep learning', in *2019 international conference on machine learning, big data, cloud and parallel computing (COMITCon)*, IEEE, pp. 72–75.

Sill, J., 1998, 'Monotonic networks', in *Proceedings of the 10th international conference on neural information processing systems*, pp. 661–667.

Section II

Big Data, Cloud and Internet of Things

7

Latest Data and Analytics Technology Trends That Will Change Business Perspectives

Kamal Gulati

CONTENTS

7.1 Introduction

This is an era of big data. These data and analytics technology trends will have significant disruptive potential over the next three to five years. Data and analytics leaders must examine the impact on business and adjust their operating, business and strategy models accordingly.

The expanded and strategic role of data and analytics in digital transformation is increasing the complexity of data, the number of variables to analyze and the types of analyses required for success. This is pushing the limits of current capabilities and approaches.

Virtually every aspect of data management, analytics content, application development and sharing of insights uses machine learning (ML) and artificial intelligence (AI) techniques to automate or augment manual tasks, analytic processes and human insight to action.

Intelligent capabilities that enable emergent and agile data fabrics, and explainable, transparent insights and AI at scale, are necessary to meet the new demands and expand adoption.

7.2 Strategic Planning Assumptions and Analysis

By 2020, augmented analytics will be a dominant driver of new purchases of analytics and business intelligence as well as data science and machine learning platforms, and of embedded analytics.

Through 2022, data management manual tasks will be reduced by 45% through the addition of machine learning and automated service-level management. By 2020, 50% of analytical queries will be generated via search, natural language processing or voice, or will be automatically generated.

By 2021, natural language processing and conversational analytics will boost analytics and business intelligence adoption from 35% of employees, to over 50%, including new classes of users, particularly front-office workers. The application of graph processing and graph databases will grow at 100% annually through 2022 to continuously accelerate data preparation and enable more-complex and adaptive data science. By 2022, 75% of new end-user solutions leveraging AI and ML techniques will be built with commercial instead of open-source platforms. By 2022, cloud-based ML services from the hyperscale cloud providers (Amazon, Google and Microsoft) will achieve the digital tipping point of a 20% share in the data science platform market.

By 2022, every personalised interaction between users and applications or devices will be adaptive.

Through 2022, custom-made data fabric designs will be deployed primarily as a static infrastructure, forcing organisations into a new wave of "cost to complete" redesigns for more dynamic data mesh approaches. By 2023, over 75% of large organisations will hire artificial intelligence specialists in behavior forensics, privacy and customer trust to reduce brand and reputation risk. By 2021, most permissioned blockchain uses will be replaced by ledger DBMS products. By 2022, more than half of major new business systems will incorporate continuous intelligence that uses real-time context data to improve decisions. By 2021, persistent memory will represent over 10% of in-memory computing memory GB consumption.

7.3 Driving Factors for Latest Data and Analytics Technology Trends

The expanded and strategic role of data and analytics in digital transformation is increasing the complexity of data, the number of variables to be analysed, the types of analysis and the speed of analysis required for success. With this increasing complexity comes ever more subtle and potentially damaging risks and challenges, such as the potential for bias and the need for transparency and trust in analytics and in ML and AI models.

The size, complexity and distributed nature of data needed for increasingly closer to real-time and optimised decision-making means that rigid architectures and tools are breaking down. This complexity is pushing the limits of current approaches, and is leading to unprecedented cycles of rapid innovation in data and analytics to meet the new requirements. At the same time, to have impact, data and analytics must be pervasive and scale across the enterprise and beyond to customers, partners and to the products themselves. The strategic technologies covered in this research in Figure 7.1 represent trends that you cannot afford to ignore. They have the potential to transform your business and will accelerate in their adoption over the next three to five years.

Many of the trends are interrelated as they are enabled by many of the same technology disruptions, but have an impact on different parts of the data and analytics technology stack (Agrawal, 2020). All have three attributes in common – they support intelligent, emergent data and analytics and are scalable for pervasive AI- and ML-driven insights and agile data-centric architectures:

- Intelligent: Advanced analytics – including AI and ML techniques – are at the core of future platforms, solutions and applications. We see the green shoots of this today. Virtually every aspect of data

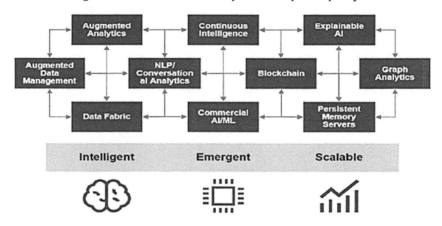

FIGURE 7.1
Latest data and analytics technology trends.

management, analytic content and application development, and sharing of insights incorporate ML and AI techniques. These are used to automate or augment manual tasks, analytic processes and human insight to action for a range of user roles (Kumar, Kumar, and Agarwal, 2019). Enhanced intelligence in all data and analytics platform components will democratise the skills required to leverage these capabilities and scale adoption across the enterprise to unprecedented levels.

- Emergent: Data and analytics is becoming more inductive in nature, assisted by AI and ML, as data models and analytics models are increasingly created by autogenerated models rather than by code. Multiple diverse structures and insights emerge *from* the data, rather than a single structure being imposed *on* the data. Intelligent capabilities that enable an emergent, agile and transparent data infrastructure are necessary to meet new demands and expand adoption. Success with data and analytics will depend on building a foundation of trust, accountability, governance and security that respects privacy and promotes digital ethics (Bhushan and Agrawal, 2020).

- Scalable: The very challenges created by digital disruption – too much data – have also created an unprecedented opportunity. Vast amounts of data are being coupled with increasingly powerful cloud processing capabilities (for both data management and data science) and the emerging capabilities enabled by a number of the top ten trends. This makes it possible to train and execute algorithms at the scale necessary to realise the full potential of AI. Realising this potential not only requires scale of processing, but also broader adoption of advanced analytics.

The compressed speed at which disruption is occurring requires data and analytics leaders to have formal mechanisms to identify technology trends, and prioritise those with the biggest potential impact on their competitive advantage. They should be added to your strategic planning, or, if part of current plans, looked at in a fresh way based on the extent to which the trends enable your top business priorities. Data and analytics leaders should proactively manage how they monitor, experiment with or deploy emerging trends. Implementing success metrics and incentives that put an emphasis on learning and reward innovation when experimenting will further contribute to success.

7.3.1 Trend 1: Augmented Analytics

Augmented analytics enables machine learning and AI-assisted data preparation, insight generation and insight explanation to augment how business people and analysts explore and analyze data in analytics and business intellgigenc (BI) platforms as shown in Figure 7.2. It also augments the expert

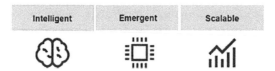

By 2020, augmented analytics will be a dominant driver of new
purchases of analytics and business intelligence as well as data
science and machine learning platforms, and of embedded
analytics.

FIGURE 7.2
Augmented analytics.

and citizen data scientists by automating many aspects of data science and
ML model development, management and deployment.

7.3.1.1 What Does It Enable?

- Augmented analytics automates aspects of finding and surfacing
  the most important insights or changes in the business (in a user's
  context) to optimise decision-making. It does this in a fraction of the
  time, with less data science and ML skills, and without prior knowledge of the relationships in data that is required when using manual
  approaches.

- Augmented analytics uses ML/AI techniques to automate key
  aspects of data science and ML/AI modeling, such as feature engineering and model selection (auto ML), as well as model operationalisation, model explanation and, ultimately, model tuning and
  management. Consequently, highly skilled data scientists have more
  time to focus on creative tasks, and on building and operationalising
  the most-relevant models. The number of models that can be tested
  increases substantially, and the time and cost to iterate and test models goes way down.

- Many autogenerated and human-augmented ML models created
  through augmented analytics are being embedded in enterprise
  applications – for example, those of the HR, finance, sales, marketing, customer service, procurement and asset management departments. This helps to optimise the elusive "last mile" of data and
  analytics – the decisions and actions of all employees, not just those
  of analysts and data scientists.

- Augmented analytics capabilities are advancing rapidly to mainstream adoption, as a key feature of data preparation, broader data
  management, modern analytics and BI as well as data science and
  ML platforms.

- Augmented analytics can also be deployed with NLP and conversational interfaces (also a top-ten trend), to allow more people across the

organisation to interact with, make predictions about and get actionable from data and insights without requiring advanced skills. It will make deep insights available to people who do not have the skills or access to ask their own questions from analytics and BI platforms.

7.3.1.2 Use Cases

Use cases span all industries and domains where data and the variables being analyzed have grown in complexity past the point of exploring completely and accurately using current approaches. Some examples include:

Banking: Before augmented analytics, banks targeted older customers for wealth management services. With augmented analytics, they found that younger clients (between the ages of 20 and 35) are actually more likely to transition into wealth management.

Agriculture: Before augmented analytics, data scientists took months to build models to find the best handful of hybrid seed combinations out of thousands to sell to farmers. With augmented analytics, domain specialist geneticists took over the process and reduced process duration to days.

Healthcare: Before augmented analytics, U.S. healthcare insurers tracked patient sickness measures as a key driver of transportation (ambulance) costs (Agrawal, 2019). With augmented analytics, they found that the main cost driver was under-12-year-olds. Investigation found that journeys were charged per person, and included parents accompanying sick children.

7.3.1.3 Recommendations

- Explore opportunities to complement existing data and analytics initiatives by piloting augmented analytics for high-value business problems currently requiring time-consuming, manual analysis.
- Build trust in machine-assisted models by fostering collaboration between expert and citizen data scientists to back-test and prove value. Understand the limitations of machine-assisted models, which work best with proven algorithms versus cutting-edge techniques.
- Monitor the augmented analytics capabilities and roadmaps of established data and analytics providers, enterprise application vendors and start-ups.
- Assess upfront setup, data preparation, openness and explainability of models, the number of variables supported, the range of algorithms provided and the accuracy of the models.
- Develop a strategy to evolve roles, responsibilities and skills, and increase investments in data literacy.

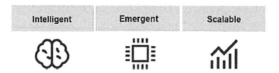

Through 2022, data management manual tasks will be reduced
by 45% through the addition of machine learning and
automated service-level management.

FIGURE 7.3
Trend 2.

7.3.2 Trend 2: Augmented Data Management

Similar to how ML and AI capabilities are transforming analytics, business intelligence and data science, across data management categories, vendors are adding ML capabilities and AI engines to make self-configuring and self-tuning processes. These processes are automating many of the manual tasks and allowing users with less technical skills to be more autonomous when using data. By doing so, highly skilled technical resources can focus on higher-value tasks, as shown in Figure 7.3. This trend is impacting all enterprise data management categories including data quality, metadata management, master data management, data integration and databases.

7.3.2.1 What Does It Enable?

- Emergent metadata can be inferred from data utilisation, users and use cases rather than descriptive metadata that is often no longer synchronised with actual data capture/write and subsequent usage.
- Active utilisation of all metadata types can be used to develop an established data fabric through processes that leverage the inventorying, cataloguing, automatic discovery of semantics, taxonomy and ontology that is crucial to data management. Organisations need to easily know what data they have, what it means, how much it delivers value, and whether it can be trusted. Utilising the statistics of existing systems – and their available capacity, known resources, cost parameters of using each resource, and the communications capability among infrastructure components – policy-level instructions will determine where data operations will take place. They can even manage their production deployment to the point of reconfiguring that deployment when necessary.
- Data fabric approaches that are composed of microservices, APIs, data pipelines and data quality/governance services, will be deployed as part of the new environment in a more customised

fashion when consistency of performance and persistence of operations is required.

- Data fusion can track which assets are used by use case, and form a knowledge and utilisation graph. When new data assets are encountered, fusion engines will analyze the similarity to other well-known data assets. They will determine their affinity to existing data/use cases, then alert other automated systems that new data is available, and is a valid candidate for inclusion (Batra and Gupta, 2016).

- Dynamic data identification allows new and existing data assets to be evaluated "in stream," and cumulative information about them to be used to develop related event models. Over time, the use cases will form processing requirements that indicate which data needs to be provided for operational and/or analytics use cases – and the rate or frequency of providing it.

7.3.2.2 How Does This Impact Your Organisation and Skills?

By 2023, AI-enabled automation in data management will reduce the need for IT specialists by 20%.

Impact on the data and analytics organisation and skills:

- Augmenting the data engineer or automating some data engineering tasks.

- Alerting data engineers to potential errors, issues and alternative interpretations of the data.

- Creating automated system responses to errors, issues and alternative interpretation of data.

- Increasing the capability to use publicly available data, partner data, open data and other assets that are currently difficult to determine as appropriate for utilisation.

- Automating data interrogation that mimics data discovery and even evaluates the "confidence" that new assets conform to known or existing models.

Impact on distributed data management:

- Continuous monitoring of capacity and utilisation for data management environments for potential redistribution of resource planning, even across multicloud and on-premises implementations.

- Optimisation and performance management that eliminates the manual (or human) determination of when to create intermediate, temporary or permanent copies of data to enhance operational or analytical performance.

- Policy-based decision engines that capture regulatory requirements as metadata configurations and then manage the hot/warm data use cases, cooler data usage, archiving expectations and even data purge or data shredding to maintain legal and audit compliance.

7.3.2.3 Use Cases

Augmented data management leverages machine learning and AI techniques for:

- Data quality: To extend profiling, cleansing, linking, identifying and semantically reconciling master data in different data sources.
- MDM: For ML-driven configuration and optimisation of record-matching and merging algorithms.
- Data integration: To simplify the integration development process, by recommending or even automating repetitive integration flows.
- DBMS: For automated management of storage, indexes, partitions, database tuning, patching, upgrading, security and configuration (Gupta and Agrawal, 2018).
- Metadata management: Well beyond data lineage, ML can be used to evaluate data rules, populate semantic frameworks and aid metadata discovery and ingestion across increasingly diverse sources (Gupta and Agrawal, 2017).

7.3.2.4 Recommendations

- Plan on fewer skilled personnel and staff with expertise in balancing physical infrastructure with logical data management requirements. Begin by exploring capacity planning tools that include dynamic hardware/infrastructure provisioning.
- Develop a strategy for isolating data assets that will be required in multicloud and on-premises scenarios for potential replication/copy instances while simultaneously considering regulatory and privacy constraints on their physical location.
- Build data literacy into the organisation. Although automation will reduce the skills barrier to using data, it will still require users to understand data and make proper uses of it. This will increase the demand for expanded metadata access and utilisation.

7.3.3 Trend 3: NLP and Conversational Analytics

NLP provides any user with an easier way to ask questions about data, as well as receive an explanation of rendered insights (Kaur and Agrawal, 2018).

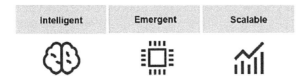

By 2020, 50% of analytical queries will be generated via search, natural language processing or voice, or will be automatically generated.

By 2021, natural language processing and conversational analytics will boost analytics and business intelligence adoption from 35% of employees, to over 50%, including new classes of users, particularly front-office workers.

FIGURE 7.4
Trend 3.

Just as search interfaces like Google made the internet accessible to everyday users and consumers, search and natural language query (NLQ) approaches to asking questions of data enable data and analytics access to mainstream business users. This includes those that don't have the skills or access to current point-and-click, visual-based analytics and BI systems, as shown in Figure 7.4.

Currently, vendors take a wide range of approaches to enabling such interfaces. Some use simple keyword matching (the familiar Google-style "search"). Others use robust natural language queries with support for more-sophisticated queries using industry-specific language. Early efforts, fore xample, may allow a user to ask for "sales by product" that generates a basic bar chart. More robust models would support questions such as, "show me top-selling products within a 50-mile radius, and comparing this year with last year." Although "top selling" is not a keyword, the platform understands that it must apply a rank function, perform geographic analytics, and then do a subquery to compare year over year.

Natural language query (NLQ) is being combined with natural language generation (NLG) to interpret and explain the results presented with automatically generated text. This text may describe visualisations on the screen, or it may describe related patterns in the data that were not specifically requested, and suggest actions in natural language. Poor data literacy is currently hindering the impact of analytics and BI initiatives. By improving consistent interpretation of insights in data, regardless of user skill, NLG (by itself and in combination with NLQ and augmented analytics) has the potential to improve the level of data literacy across an organisation. Here too, the textual descriptions may be in the form of written text, or voice-generated, or both. The ability to control aspects of language such as verbosity and tone – and how much of the text is form-based, template-driven or automated – varies between products.

Conversational analytics is still emerging, but takes the concept of NLQ and NLG a step further, first by allowing such questions to be posed verbally. This voice interaction could be through a digital assistant (such as Amazon Alexa), on a mobile phone, or using other devices. In addition to voice support for queries, these capabilities are emerging to be conversational in the form of a virtual AI assistant or bots. NLP/conversational interfaces will also leverage insights and embedded ML/AI models generated from augmented analytics. The virtual assistant may answer the initial question posed, but then elaborate with autogenerated insights from augmented analytics, and suggestions in natural language such as, "there was a spike in sales for purple products in week 30." It will also make predictions and prescribe actions.

7.3.3.1 What Does It Enable?

- Most analytics and BI tools currently require users to choose data elements and place them on a page to create queries and visual analysis. NLP/conversational analytics brings ease of use to a new level, and allows a query to be as easy as a Google-like search or conversation digital assistants, such as Alexa.

- Any user can ask questions using text or voice with increasingly complex questions and responses. NLP is increasingly an interface to querying and interacting with autogenerated insights from augmented analytics.

- The combination of NLP with augmented analytics – including automatic insight generation – allows users to rapidly find the proverbial needle in a haystack, presenting the most-important and actionable insights via conversational analytics with natural language generation.

- More-robust NLP interfaces include industry or domain-specific taxonomies and linguistics (such as HR, finance, healthcare or financial services) so that phrases are more correctly interpreted.

- NLP and conversational interfaces are being embedded in analytics and BI platforms, digital personal assistants, chatbots and applications.

- NLP provides intuitive forms of communication between humans and systems. NLP includes computational linguistic techniques (both traditional and machine learning) aimed at recognising, parsing, interpreting and generating natural languages.

- NLP and underlying knowledge graphs are being extended to analyze unstructured and other data types, and used as a foundation for data science as well as ML/AI models.

7.3.3.2 How Does This Impact Your Organisation and Skills?

Conversational analytics can dramatically improve the adoption of analytics by every employee, rather than by predominant power users and business analysts, resulting in higher business impact. However, as access to more powerful insights permeates the organisation at all skill levels, there is a need for a formal and enterprise wide focus on improving the data literacy of all users (Pathak and Agrawal, 2019).

7.3.3.3 Use Cases

Instead of logging into a complicated dashboard, any user – from the C-suite to analysts to operational workers – can interact verbally with a virtual personal assistant or their mobile phone, and ask for an analysis that is relevant to them. In combination with augmented analytics capabilities, a salesperson might, for example, ask for an analysis of sales or a pipeline. Or the system may have learned that the sales manager also looks at this information. Based on that person's role and/or behavior, they will be served up an explanation or narrative – in text or voice – of statistically important drivers of change. They could then be sent visualisations (via a device) that show important trends, patterns or outliers based on their role.

Predictions and prescriptive recommendations could also be communicated to the sales manager. Conversational analytics will also be embedded in the workflow of applications that every employee uses. The starting point of the analysis might be an autogenerated visualisation or insight from augmented analytics where any user can explore further by asking additional questions in text or voice.

7.3.3.4 Recommendations

- Evaluate the capabilities, roadmaps and partnerships of your analytics and BI platform and enterprise application vendors, as well as those from start-ups.
- Assess the maturity and scalability of solutions, particularly in terms of integration and ease of use, upfront setup/configuration requirements, languages supported and types of analysis.
- Invest in data literacy as a critical element of success.

7.3.4 Trend 4: Graph Analytics

Graph analytics is a set of analytic techniques that allows for the exploration of relationships between entities of interest such as organisations, people and transactions. One application of graph analytics – graph-enabled semantic knowledge graphs – forms the foundation of many NLP/conversational

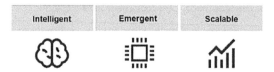

The application of graph processing and graph databases will grow at 100% annually through 2022 to continuously accelerate data preparation and enable more-complex and adaptive data science.

FIGURE 7.5
Trend 4.

interfaces and data fabrics, and enriches and accelerates data preparation and data science (see Figure 7.5).

7.3.4.1 What Does It Enable?

Graph analysis shows how tightly several trends or data points are related to each other. For example, an automated lighting system could be set to come on at dusk, but it might also come on during a severe weather situation, and time-of-day and seasonal variations may be combined with "waking hours" evaluation models. All of these inputs may or may not be related to each other, and graph analytics shows these as compact (in terms of their degree of relatedness) "nodes" of data. It also reveals whether those nodes are tightly compacted or just loose connections. When conditions change, the demarcation points between nodes may change, and a whole new set of nodes may dynamically emerge. Graph analysis also shows if something is only a correlation, or if the two (or more) things are related as precedents with dependency, or just have "causal" relationships. This means the nodes are connected explicitly or implicitly, indicating the levels of influence, frequency of interaction or probability.

Graph models determine "connectedness" across data points and create clusters based on levels of influence, frequency of interaction and probability. Once highly complex models are developed and trained, the output is easier to store because of the expanded capabilities, computational power and adoption of graph databases. The user can interact directly with the graph elements to find insights, and the analytic results, and output can also be stored for repeated use in a graph database.

Graph databases therefore present an ideal framework for storing, manipulating and analyzing graph models, although some graph analytics vendors have their own graph analytics engines that do not require a separate graph database. Generating a dynamic graph about how different entities of interest – people, places and things – are related, instead of more-static relational schemes, enables deeper insights that are closer to human knowledge representation. For example, it can easily combine, relate and find dynamic insights from data from exercise apps, diet planners, medical records and health news feeds.

Graph technology underpins the creation of richer semantic graphs or knowledge bases that can enhance augmented analytics models, as well as the richness of conversational analytics. Graph analytics also supports the creation of emergent metadata management and data catalogues. It does so by capturing all of the knowledge about what data you have, where it resides, how it is all related, who uses it, why, when and how. That insight can be leveraged by ML models to make recommendations to provide more-personalised, automated and properly governed insights to the business and its applications.

7.3.4.2 How Does This Impact Your Organisation and Skills?

- New skills will be required. These include graph-specific standards, graph databases, technologies and languages such as Resource Description Framework (RDF), SPARQL Protocoland RDF Query Language (SPARQL); and emerging languages, such as Apache TinkerPop or the recently open-sourced Cypher. The scarcity of these skills has been an inhibitor to graph adoption.
- Commercialisation of graph analytics is still at an early stage, with a small number of emerging players.
- While current technologies still require specialist skills, SQL-to-graph interpreters are emerging. These interpreters convert graph-based procedures into disaggregated procedural SQL (and back again). These capabilities are helping to make graph technologies compatible with existing atasets.

7.3.4.3 Use Cases

The number of use cases is increasing with the need for complex analysis. They range from fraud detection to customer influencer networks, through to social networks and semantic knowledge graphs. Conversational analytics, health advisors, financial crimes and risk detection can also leverage graph capabilities.

Other specialised applications include:

- Route optimisation between waypoints for transportation, distribution and even foot traffic.
- Market basket analysis to show products that have a coincidental or dependency relationship.
- Fraud detection to identify clusters of activity around connected groups of "actors."

- Social network analysis to determine the presence of influencers, "canaries," decision-makers and dissuaders.
- CRM optimisation to determine probabilities of success for things like "next best offer."
- Location intelligence to determine probable routes or locations from known data points when location is otherwise indeterminate.
- Load balancing in any type of network system like communications or utilities, but also something more concise like computer networks.
- Special forms of workforce analytics, such as enterprise social graphs and digital workplace graphs.
- Recency, frequency, monetary analysis of related networks of objects, assets and conditions to help designate the best resources at the best utilisation point.
- Law enforcement investigation to isolate missing or unknown risks and even identities (such as third-party relationships in child endangerment cases).
- Epidemiology for analyzing the intersection of environment, overall condition of a patient, diet, exercise, acute/chronic disease analysis, interaction of therapeutic and chemical/pharmaceutical treatment regimens.
- Genome research for gene interaction to determine the potential for and treatment of existing hereditary disease conditions to assist with targeted medical practices.
- Detection of money laundering to unearth relationships among "actors" to identify malignant versus benign actions.

Data and analytics leaders should:

- Evaluate opportunities to incorporate graph analytics into their analytics portfolios and strategies for both analytic applications, and as part of the underlying data fabric in the form of semantic knowledge graphs.
- Assess the products available from both incumbent and new vendors, but recognise that most solutions are maturing, and are often focused on specific domains and verticals.
- Explore the use of new graph-to-SQL interpreters, but also invest in developing the necessary unique skills and competencies.

7.3.5 Trend 5: Commercial AI/ML Will Dominate the Market over Open Source

Open-source libraries and development environments have provided a much-needed democratisation of the data science and machine learning fields, as well as innovation and flexibility.

But, as was the case in the past with predictive analytics techniques, flexibility and power can come at the expense of disciplined operationalisation mechanisms. Many models have been created, but their business value has not been fully realised as they have not been put into production at scale. Data science and machine learning teams are now starting to be measured on business results rather than production metrics (the number of models produced, or projects started, for example). Consequently, the required disciplined approach brought about by commercial platforms is becoming a required condition to achieving business value and data science team sustainability.

7.3.5.1 What Does It Enable?

- Better productivity: The assembly and retro-fitting of diverse OS tools requires lots of skills and manual labor, which has become the focal point of commercial attention, along with a clearer path toward business value.
- Democratisation of AI: Commercial providers will increasingly "smooth" the rough edges often associated with open-source projects by orchestrating the user experience and "connecting the dots." This provides much more integrated capabilities through orchestration (see "Embracing Competition to Evolve and Enable Data and Analytics Product Offerings"). This also makes cutting-edge AI/ML development much more available to the broader skill set found in most enterprises.
- Better AI/ML planning and roadmaps: Current AI strategies are full of uncertainty (Agrawal, 2016), conflict and vagueness. The comeback of capable commercial platforms will increase plannability and IT roadmaps by providing concrete anchor points into the IT software infrastructure.

7.3.5.2 How Does This Impact Your Organisation and Skills?

Increased use of commercial data science and machine learning will help to narrow the current skills gap in these areas. It will also facilitate greater collaboration among AI/ML developers with varying levels of skill.

7.3.5.3 Use Cases

Marketplaces for algorithms and pretrained models from Amazon Sage Maker, Google AI Hub, Microsoft, KNIME, RapidMiner will be accelerators of commercial data science and ML platforms. This will further extenuate the need for building a core competency around collecting data, managing it, curating it and making it accessible in an agile way.

7.3.5.4 Recommendations

- Start or continue to upskill existing professional staff (from business and technology departments, for example) into a cast of citizen data scientists, while progressively moving to a commercial production environment.
- Emphasise easy-to-use augmented data science/ML solutions and upcoming marketplaces as catalysts for simplified ML adoption, or to bootstrap ML efforts.
- As your data science, ML and AI capabilities mature and scale, instigate success metrics for the models you have deployed into production, and the business impact of those models, rather than the number of models created.
- Create an open-source audit process aimed at validating open-source models before integrating them into a commercial production environment.
- Focus on data management competency because, as algorithms commoditise, data will be the critical determinant of AI/ML success.

7.3.6 Trend 6: Data Fabric

Deriving value from analytics investments depends on having an agile and trusted data fabric. A data fabric is generally a custom-made design that provides reusable data services, pipelines, semantic tiers or APIs via combination of data integration approaches (bulk/batch, message queue, virtualised, streams, events, replication or synchronisation) in an orchestrated fashion. Data fabrics can be improved by adding dynamic schema recognition, or even cost-based optimisation approaches (and other augmented data management capabilities). As a data fabric becomes increasingly dynamic, or even introduces ML capabilities, it evolves from a data fabric into a data mesh network. Figure 7.6 represents Trend 6.

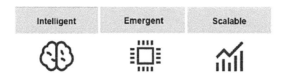

| Intelligent | Emergent | Scalable |

Through 2022, custom-made data fabric designs will be deployed primarily as a static infrastructure, forcing organizations into a new wave of "cost to complete" redesigns for more-dynamic data mesh approaches.

FIGURE 7.6
Trend 6.

7.3.6.1 What Does It Enable?

The best approach to understanding the intersection of data integration strategies is to start from traditional practices, and then simultaneously consider new or emerging practices in architecture and infrastructure design. This will allow you to compare and contrast these practices with your organisation's aspirational design goals.

Data integration approaches now (and in the future) must effectively address the same three issues:

1. Metadata drives the overall comprehension and performance optimisation of any data asset in the organisation.
2. Perpetual connections are not yet above 80% reliability when it comes to periodically connected devices or sensors at the edge, and lack the physical infrastructure needed for reliable communications.
3. Processing is always a mobile consideration and, as a logical requirement, can be relocated in such a way that the process can be brought to the data or the data to the process. This results in the negation of most considerations for selecting the deployment platform (including cloud, on-premises, multicloud and hybrid, among other combinations).

Current and traditional practices are focused on forms of physical consolidation combined with semantic interpretation:

- Data warehouses, data lakes and operational data stores remain prominent as data repositories that represent specific persistence and performance expectations.
- More-traditional data integration approaches still prefer to identify and design specific targets. For comprehension and performance reasons, the data warehouse, base operational data store systems, and even the logical data warehouse use a combination of the physical data lake and the physical data warehouse alongside some type of unifying semantic access tier.
- Data hubs represent a next step in traditional design. They combine the idea of distributed data management for geographic or domain "regions" with a data services back-plane that exchanges data in predispositions developed using microservices, APIs or P2P data services in the form of desktop as a service (DaaS).

If the metadata, physical management and processing design are considered as the primary "poles" for deployment, the infrastructure design has two complementary approaches:

- Data fabric is more of a designed approach, mostly tending toward use cases and locations on either "end" of a thread. The threads may cross and do handoffs in the middle, or even reuse their parts, but they are not built up dynamically. They are merely highly reusable, normalised services.

- Data mesh is a fully metadata-driven approach. Statistics in the form of metadata accumulation are kept relating to the rate of data access; platform, user and use case access; the physical capacity of the system; and the utilisation of the infrastructure components. Other data points include the reliability of the infrastructure, the trending of data usage by domain and use case, and the qualification, enrichment and integrity (both declared and implied) of the data.

7.3.6.2 How Does This Impact Your Organisation and Skills?

Specific roles will undergo new workflows and definitive changes in their responsibilities and tasks.

- Data engineer: This role is composed of tasks that determine the accessibility, qualification, delivery and processing of data. It requires an understanding of how data is used in a broad spectrum of business processes, the variable semantic and schema-based approaches to using the same data concepts, and the widely varied data quality presented when data is combined.

  The human capabilities of data engineers will be augmented by AI/ML processes that identify nearly all of the initial pain points for data refactoring, modeling, schema production and data quality recognition. These processes might even provide expert advice regarding infrastructure decisions – if not providing actual, dynamic resource allocation and provisioning.

- Data scientist: This role will benefit from data fusion outputs that create alerts about expanding data assets. These alerts will be specifically tuned for the current project – but, can also begin to recognise data that a given scientist utilises in terms of data patterns.

- Data modeler: Data modelers, data integration developers and database administrators responsible for data modelling will model less, and verify more. Data models are always present, some formal and others "not so much." In less-structured data, models can be collaborative.

- Information architect: Information architects determine the alignment of intended functions of information collection and management with the appropriate form needed for acquiring, managing, accessing and utilising data. Information architects working with data fabric will need to focus on identifying the required functionality of a data asset and imputing it as metadata.

7.3.6.3 Use Cases

- Dynamic data engineering: The data fabric architecture can be investigated and adopted for more-dynamic, reusable and optimised data engineering pipelines. This can be designed once, and can then work in "auto" execution mode, with data engineers acting as adjudicators and facilitators rather than developers of integrated data flows. They need to be flexible to changing data environments (cloud, multicloud and hybrid cloud) and even the users and use cases involved in this data.
- Governed/trusted data science: Data scientists now demand end-to-end lineage and understanding of their data models and algorithms for efficiency and compliance, and the data fabric design helps them with data fusion outputs that alert them to expanding data assets. This gives them more visibility into their active metadata (including performance optimisation, data quality, design and lineage, among several others) and allows them to be in more control of their projects.
- **Logical data warehouse architecture:** The data fabric and mesh concepts will liberate data management from having to confine data to a physical consolidation combined with static semantic interpretation. This limits reuse of integrated data across heterogeneous applications needing specific data models and semantic interpretations.

7.3.6.4 Recommendations

- Begin tracking the origins and sources of data as well as types of reporting and analysis based on the characteristics of all of the use cases.
- Introduce data fabric designs when the primary data management and integration approach is focused on *connecting* (as opposed to *collecting*) data. Many of the tools used to build a data fabric are also referred to as "semantic tools," and these tools can also move data to collections as write-out caching tiers.
- Apply a three-tiered data fabric design that isolates industrywide data for one design and one common conceptual model, the organisation's prevalent (but not universal data in a second tier, and transactional values as an almost "attached data only" third tier.
- Build a pilot data asset resource utilisation engine that uses machine learning-enabled data catalogs. Start by collecting the required metadata and performing analysis over this meta data to determine the specifications for a simple "alert" function that tells users that new data is available – possibly from a data lake.

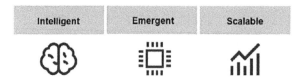

By 2023, over 75% of large organizations will hire artificial intelligence specialists in behavior forensic, privacy and customer trust to reduce brand and reputation risk.

FIGURE 7.7
Trend 7.

7.3.7 Trend 7: Explainable AI

As we move toward more augmented analytics including autogenerated insights and models, the explainability of these insights and models will become critical to trust, to regulatory compliance, and to brand reputation management. Explainable AI is the set of capabilities that describes a model, highlights its strengths and weaknesses, predicts its likely behaviour, and identifies any potential biases. It has the ability to articulate the decisions of a descriptive, predictive or prescriptive model to enable accuracy, fairness, accountability, stability and transparency in algorithmic decision-making. Figure 7.7 represents Trend 7.

7.3.7.1 What Does It Enable?

Explainable AI enables a better adoption of AI by increasing the transparency and trustworthiness of AI solutions and outcomes. Explainable AI also reduces the risks associated with regulatory and reputational accountability for safety and fairness. Increasingly, these solutions are not only showing data scientists the input and the output of a model, but are also explaining the reasons the system selected particular models and the techniques applied by augmented data science and ML. Bias has been a long-standing risk in training AI models. Bias could be based on race, gender, age or location. There is also temporal bias, bias toward a specific structure of data, or even bias in selecting a problem to solve. Explainable AI solutions are beginning to identify these and other potential sources of bias.

Explainable AI technologies may also identify privacy violation risk, with options for privacy-aware machine learning (PAML), multiparty computation and variants of homomorphic encryption to identify privacy violation risks.

7.3.7.2 How Does This Impact Your Organisation and Skills?

Diversity is a critical foundation for explainable AI because:

- Diversity of data is necessary to have an objective view of the problem and deliver trusted outcomes.

- Diversity of algorithms is necessary to make trade-offs, especially to resolve the dilemma of accuracy versus explainability.
- Diversity of people is necessary to build a successful AI team and an AI ethics board to minimise reputational and business risks, as well as to ensure AI safety.
- Data and analytics leaders should invest in training and education to develop the skills needed to mitigate risks in black-box models.

This should include:

- How to make data science and ML models interpretable by design, and how to select the right model transparency from a range of models, from least to most transparent (taking into account their implications for accuracy).
- How to select the right model accuracy when required, and methods of validating and explaining these models "post hoc," such as model-agnostic or model-specific explainability; global (across the entire model) or local (a specific output) explainability.

- Various methods, such as generative explainability and combining simple, but explainable models, with more complex, but less explainable ones.
- Exploring the latest explainability techniques, such as the ones that are tracked by DARPA, or that are coming from commercial vendors.
- Visualisation approaches for seeing and understanding the data in the context of training and interpreting machine learning algorithms (for example, visualisation of correlations or of outliers in the data).
- Techniques for understanding and validating the most-complex types of predictive models, such as sensitivity analysis, surrogate models and "leave one covariate out" (LOCO).
- Communication and empathy skills for data scientists to detect the users' attitude and needs for explainability and successful AI adoption
- Establish AI ethics boards and other groups that are responsible for AI safety, fairness and ethics. These boards should include internal and external individuals known for their high reputation and integrity.

7.3.7.3 Use Cases

The commercial sector delivers explainable AI to analytics and BI platforms and in data science and ML platforms in various ways.

Commercial offerings include:

- Data science platforms such as Data Robot Labs and H2O.ai, which automatically generate model explanations in natural language.
- tazi.ai provides pattern visualisations (such as profit/loss prediction patterns) for nontechnical business users and enables them to investigate possible patterns with explanations in an interactive fashion.
- Darwin AI offers Generative Synthesis technology, which provides a tool for granular insights into neural network performance.
- Zest Finance specialises in the financial industry, and provides accurate and transparent Auto ML models for lending.
- Salesforce Einstein Discovery explains model findings and will alert users to potential bias in data.

7.3.7.4 Recommendations

Data and analytics leaders should assign responsibility and focus on the following aspects of AI explainability:

- Include an assessment of AI explainability features when assessing analytics, business intelligence, data science and machine learning platforms.
- Use the features provided by selected vendors to build trust and expand your adoption of autogenerated models and insights.
- Assess the trade-off between accuracy and explainability on a use-case-by-use-case basis.
- Develop governance approaches that decide when explainability is necessary, and guidelines that assess the trade-offs.
- Understand that not all AI solutions require explainability, but those that affect user adoption, regulation and risk probably will.
- Establish accountability for determining and implementing the levels of trust and transparency of data, algorithms and output for each use case.

7.3.8 Trend 8: Blockchain in Data and Analytics

The promise of blockchain is substantial. It offers cryptographically supported data immutability, shared across a network of participants. But this comes with massive complexity. Your internal business processes effectively become shared across your network. And your blockchain-based system won't be your system of record, meaning a huge integration effort involving data, applications and business processes. Lastly, the technology hasn't matured to real-world, production-level scalability yet (see Figure 7.8).

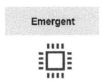

Emergent

By 2021, most permissioned blockchain uses will be replaced by ledger DBMS products.

FIGURE 7.8
Trend 8.

7.3.8.1 What Does It Enable?

Block chain technologies address two challenges in data and analytics. First, blockchain provides the full lineage of assets and transactions. Second, blockchain provides transparency for complex networks of participants.

A shared ledger exposes an entire participant network as a graph of interacting, connected nodes. This makes it possible to identify:

- Which participants are interacting with each other.
- How a first interaction potentially triggers second or third interactions.
- How interactions change over time, and whether they are persistent or transient.
- Who the most important (and possibly most influential) network participants are.

7.3.8.2 How Does This Impact Your Organisation and Skills?

Organisations adopting blockchain will need to recast their existing centralised business processes in the context of a distributed computing environment. Traditional skills around data, application and process integration will be essential to successful blockchain evolution. Enterprises will also need to develop blockchain-specific application development skills, including smart contracts. Complicating this work are challenges around the lack of interoperability between competing blockchain implementations. Until the blockchain technology landscape becomes less volatile, organisations should expect to support multiple blockchain implementations.

7.3.8.3 Use Cases

Auditing and product lineage: An irrevocable dataset shared across all supply chain participants also extends visibility throughout the network. This increases the amount of surface area that can be monitored, improving traceability and understanding of product provenance. Early work in this area is being pioneered by firms like Ever ledger, which traces the global movement of diamonds and other expensive assets, and Provenance, which targets the supply of things like fish from Indonesia to Japan.

Fraud analytics: A public distributed ledger potentially enhances fraud analytics and risk management by providing all parties with the same set of data. Fraudulent activity detected by one party could be quickly flagged and distributed throughout the network, notifying other participants before they become victims. Any industry involving multiparty transactions or coordination, like healthcare or insurance, will likely benefit from the fraud analytics possibilities available from blockchains or distributed ledgers.

Data sharing and collaboration: When external data is brought into an internal process, it must be treated as completely unverified and untrusted data. This is a reasonable (if somewhat paranoid) approach when data originates from a foreign governance model. The content and context of the data is outside your control, impacting how data is described, shared and used. The guarantees of irrevocability and consistency provided by blockchain implementations, and supported by a shared or self-describing data model, allow you to treat external data as if it originated internally. Data creation and modification can be tracked to its source, offering guarantees on data lineage and provenance.

7.3.8.4 Recommendations

- Position blockchain technologies as supplementary to your existing data management infrastructure by highlighting the capabilities mismatch between data management infrastructure and blockchain technologies.
- Explore unique use cases in partnership with business leaders by forming a pilot team to exploit the strengths of blockchain in areas of immutability, fault tolerance and transactional transparency.
- Design for flexibility. Given the early, volatile state of blockchain technologies, limit your long-term commitments by introducing abstraction layers around vendor-specific platforms when possible.

7.3.9 Trend 9: Continuous Intelligence

Continuous intelligence combines data and analytics with transactional business processes and other real-time interactions. It leverages augmented analytics, event stream processing, optimisation, business rule management and ML (see Figure 7.9).

7.3.9.1 What Does It Enable?

Continuous intelligence is the combination of three things:

1. Situation awareness is real-time or near-real-time, based on continuous (always on) ingestion of streaming data.

Intelligent   Scalable

By 2022, more than half of major new business systems will incorporate continuous intelligence that uses real-time context data to improve decisions.

FIGURE 7.9
Trend 9.

2. Proactive behavior means that the system can push alerts, update dashboards or trigger automatic responses when it detects a situation that requires attention (it's not just reactive, waiting for a person or application to inquire or pull information).

3. Prescribing behavior means that the system tells you what to do. It provides decision support for human decisions or decision automation for lights-out processes (in other words, it's not just notifying you of what's happening and making you figure out an appropriate response). Continuous intelligence improves the quality and precision of a wide variety of operational decisions because it incorporates more kinds of trusted data into the algorithms that are used to compute decisions. It is relevant to real-time and near-real-time decisions where there is a benefit to having an understanding of the current situation (or events in the past few seconds or minutes).

Systems are able to process high volumes of data quickly, shielding people from overload. They are able to apply rules and optimisation logic to evaluate far more options than a person could consider in the available time.

7.3.9.2 How Does This Impact Your Organisation and Skills?

Continuous intelligence has a bigger impact on data and analytics teams than most of the other trends outlined here because it directly affects transaction processing, customer-facing, logistics, B2B and other operational systems. Data and analytics teams must work hand in hand with application architects, application developers, business process management analysts, and business analysts to design, build, deploy and maintain the systems.

7.3.9.3 Use Cases

Rudimentary, stand-alone versions of continuous intelligence are already ubiquitous. The navigation system in your mobile device is a continuous intelligence system that provides real-time advice on what route to take.

In general, there are two types of continuous intelligence systems:

- Proactive push systems: People are overloaded with information, so they need management by exception. Always-on, continuous intelligence (monitoring) systems run all day, listening to events as they occur. When they detect a threat or opportunity situation that requires a response, they update dashboards, send alerts, or trigger an automated response. These are proactive – a push – because the system initiates the response. They might kick off a process or send a control signal to a machine. Real-time dashboards provide situation awareness.
- On-demand: All continuous intelligence systems are proactive in some of their processes by definition. However, virtually all are also reactive (on-demand) for some other processes. On demand, real-time analytics are invoked when a business process reaches a decision point, or when a person chooses to trigger a process or decision (such as a loan approval or a next best offer).

Major business benefits are emerging from corporate usage scenarios of continuous intelligence including:

- Turning 360-degree views of customers into real-time 360-degree views for more precise and effective customer offers and improved customer support.
- Implementing condition-based, predictive maintenance to extend the life of machines while reducing wasteful, premature replacement of parts.
- Providing enterprise nervous systems for airlines, railroads, trucking companies and shipping fleets that monitor and optimise resource scheduling decisions while improving customer satisfaction.

7.3.9.4 Recommendations

Data and analytics leaders should:

Work collaboratively with application architects, application developers, business process management analysts and business analysts to develop a centralised plan and design, build, implement and maintain the systems.

7.3.10 Trend 10: Persistent Memory Servers

See Figure 7.10 for Trend 10.

7.3.10.1 What Does It Enable?

The amount of data available and required for modern systems (across all areas of IT) is growing rapidly and the urgency of transforming data into

Scalable

By 2021, persistent memory will represent over 10% of in-memory computing memory GB consumption.

FIGURE 7.10
Trend 10.

value in real time is growing at an equally rapid pace. New server workloads are demanding not just faster CPU performance, but massive memory and faster storage. Historically, DRAM has been the costly but reliable byte-addressable memory solution, but it still lacks the economics of the much cheaper and denser (but slower) non-volatile NAND flash memory used as block-addressable storage. This new persistent NVDIMM does not replace DRAM, but is used together to optimise software environments.

Intel DV Optane persistent memory has two modes of operation:

1. Memory mode: Plug and play as a larger memory pool.
2. Application direct mode: Persistent memory pool, requiring application modification and optimisation.

In memory mode, large memory sizes are available with no changes necessary to the software. However, memory mode does not include persistence. Although far easier to use (as no modifications are necessary) it loses the benefits of persistence for high availability (HA), as it remains volatile.

In application direct mode, large memory sizes are also available and are persistent, realising the true benefits of persistent memory. This mode requires vendors to modify their software to take advantage of persistent memory. This will have a profound effect on HA, disaster recovery and how a system restarts after a failure.

7.3.10.2 How Does This Impact Your Organisation and Skills?

There is little or no change to the skills required to use persistent memory as necessary changes are within the DBMS software. Over time, as the cost drops, and as servers with persistent memory become widely available, the use of in-memory DBMS will grow. The impact on the organisation will be in areas of consolidation, more-efficient HA for in-memory systems, higher performance, and the enabling of new architectures, such as hybrid transactional/analytical processing (HTAP).

7.3.10.3 Use Cases

The use cases for persistent memory are many and varied, including:

- Virtualisation: Most software virtualisation (such as VMware) requires memory. Persistent memory will allow larger, more-efficient virtualisation environments. This will be true not only for server virtualisation, but also for desktop virtualisation systems.
- DBMSs and data grids: Data grids are already in-memory architectures, and DBMSs are increasingly using in-memory techniques. Persistent memory will allow most or all data to be in-memory, not only boosting performance but also simplifying HA while reducing the restart implications for HA systems.
- Analytics: Many analytics vendors today make use of in-memory structures to increase performance. With persistent memory, these systems will be able to increase the amount of data to keep in memory, and therefore increase the performance even more. This will be especially true of augmented analytics using ML algorithms where massive amounts of data are required.

7.3.10.4 Recommendations

Data and analytics leaders interested in data management solutions should:

- Ask your software providers to identify what they are doing to take advantage of persistent memory, and when it will be available.
- Evaluate the use of persistent memory servers (as they become available) in your infrastructure for simplification of HA architecture, increased use of virtualisation, server consolidation and potential cost savings.

References

Agrawal, R., 2016, 'Design and development of data classification methodology for uncertain data', *Indian Journal of Science and Technology*, vol. 9, no. 3, pp. 1–12.
Agrawal, R., 2019, 'Predictive analysis of breast cancer using machine learning techniques', *Ingeniería Solidaria*, vol. 15, no. 29, pp. 1–23.
Agrawal, R., 2020, 'Technologies for handling big data', in *Handbook of research on big data clustering and machine learning*, IGI Global, Hershey, PA pp. 34–49.
Batra, M. and Gupta, N., 2016, 'Various security issues and their remedies in cloud computing', *International Journal of Advanced Engineering, Management and Science (IJAEMS)*, vol. 2, no. 2, pp. 18–20.
Bhushan, D. and Agrawal, R., 2020, 'Security challenges for designing wearable and IoT solutions', in *A handbook of internet of things in biomedical and cyber physical system*, Springer, Cham, pp. 109–138.
Gupta, N. and Agrawal, R., 2017, 'Challenges and security issues of distributed databases', in *NoSQL*, Chapman and Hall/CRC Boca Raton, FL, pp. 265–284.

Gupta, N. and Agrawal, R., 2018, 'NoSQL security', in *Advances in Computers*, Vol. 109, Elsevier, pp. 101–132.

Kaur, S. and Agrawal, R., 2018, 'A detailed analysis of core NLP for information extraction', *International Journal of Machine Learning and Networked Collaborative Engineering*, vol. 1, no. 01, pp. 33–47.

Kumar, A., Kumar, P. S. and Agarwal, R., 2019, March, 'A face recognition method in the IoT for security appliances in smart homes, offices and cities', in *2019 3rd international conference on computing methodologies and communication (ICCMC)*, IEEE, pp. 964–968.

Pathak, S. and Agrawal, R., 2019, 'Design of knowledge based analytical model for organizational excellence', *International Journal of Knowledge-Based Organizations (IJKBO)*, vol. 9, no. 1, pp. 12–25.

8

A Proposal Based on Discrete Events for Improvement of the Transmission Channels in Cloud Environments and Big Data

Reinaldo Padilha França, Yuzo Iano, Ana Carolina Borges Monteiro, Rangel Arthur and Vania V. Estrela

CONTENTS

8.1 Introduction

Big data refers to storing a huge amount of data, as well as the ability to derive value from that information at a rapid rate. This technology is based on 5 Vs, which are value, volume, velocity, variety and veracity. In this sense, value is the investment required to generate returns for companies, which can improve service quality and increased revenue. Volume refers to the fact that big data is based on the huge volume of data generated each day; velocity because the way this information is managed must be dynamic, otherwise, it loses its value. Variety refers to the fact that data originates from a multitude of distinct channels, including email, social media, sensors, and many others; and finally, veracity because if the data is not real it is of no use (Oussous et al., 2018)

Big data refers to the sheer volume, variety and speed of data that demand innovative and cost-effective ways of processing information for better insight and decision-making. Big data has a recent history, but many of the foundations on which it was built are long-established. In recent years the evolution of big data has accelerated further, following the growth of data, made available both by the advancement of digital technology and processing software, as well as new consumer habits, encouraged by the rise of e-commerce, applications and resources, geolocation and intensive use of social networks. Big data can be described in terms of data-management challenges that due to increasing data volume, speed and variety cannot be solved with traditional databases (Grover et al., 2018)

Big data was born in the 1990s, with the use of the term "big data" when NASA (National Aeronautics and Space Administration) wanted to describe large complex data sets that are not always easy to translate with existing technology. However, when manipulated, these data become a region of transversal knowledge that encompasses various productive fields and scientific research. At the time, "big data" was used to describe complex data sets that challenged the traditional computational boundaries of capture, processing, analysis and information storage. And in 2001, then vice president and research director of Enterprise Analytics Strategies Doug Laney articulated the definition of big data in the three V's: volume, variety, and velocity, which have now been expanded (Zomaya and Sakr, 2017).

Because of its efficiency, organisations have begun to realise the power of using big data, impacting the way new business has sprung up and rocketed, but also because some giants in the corporate world have come into play and now invest in data processing and analysis. As data to drive decisions because of its powerful insights, big data impacts on business and society as a whole. Despite all the excitement, many organisations don't realise they have a big data problem; in general, an organisation is likely to benefit from big data technologies when existing databases and applications can no longer expand into big data (Erevelles et al., 2016; Loebbecke and Picot, 2015).

Cloud computing is the on-demand delivery of computing power, database storage, applications and other IT resources over the Internet at pricing based on usage. When it comes to cloud computing, we talk about the ability to access files and perform different tasks over the Internet, i.e., it is not necessary to install applications on the computer, since it is possible to access different online services to perform the given task, since data is not on a specific computer but on a network. Once properly connected to the online service, you can enjoy your tools and save all the work done to access it from anywhere, precisely because the operating "computer" will be in the cloud, and because access will be from any computer that has access to the Internet (Rittinghouse and Ransome, 2017).

The key features of cloud computing technology are agility, scalability, access from anywhere and across different devices (mobile phones, notebooks and mobile devices), allowing sharing of resources by a large group of users, easy-to-use services and no installation required. However, the security issue is something of a concern. Overall, cloud computing delivers faster innovations, flexible capabilities and economies of scale, where you get paid for just what you use, reducing operating costs, running processes more efficiently and scaling business as needed. To survive and thrive in a volatile, competitive, technology-driven market, companies need to be in tune with the latest management practices. Cloud computing is a technology that enables greater agility, flexibility and cost savings in a company's information technology infrastructure (Yang et al., 2017)

This study aims to implement a discrete event-based model to improve the transmission of information in communication systems, through the pre-coding process of bits applying discrete events in the signal before the modulation process, using advanced modulation format Differential Quadrature Phase Shift Keying (DQPSK) with Rician Fading, which can be applied to cloud environments and big data in environments and scenarios with CPS application, improving transmission of data. The concepts that contribute to the development of the proposed methodology will be presented and discussed.

8.2 Big Data

Big data relates to a large volume of data that is organised or not, in an orderly manner. In other words, it refers to the storage and processing capacity of an immeasurable number of distinct information. Data analysis is already part of the business routine of many areas of business, and is a strategy that helps streamline processes and understand customer and market behavior patterns to make services and products more profitable. But the amount of information available has never been greater, and it is becoming more unfeasible to perform this task manually every day. Thus, big data is a technology that enables high performance and high availability information processing, with digital tools that make data collection, processing and visualisation simpler, standardised and effective. The complete concept is based on five principles and each of them will have a direct influence on the performance of solutions available in the market. These principles are known as the Big Data 5 Vs (Bihl et al., 2016).

Volume refers to the amount of data, measured in Giga, Tera or even Zettabytes. With the advancement of technology, storage capacity has increased and provided an ever-increasing record of information. A big data tool must be able to handle a large amount of data, much of which is due to social networks, smartphones, mobile internet and IoT devices. The amount of information that circulates in digital media grows continuously. That is why we are increasingly dependent on big data tools, which through artificial intelligence and machine learning have led us to a new pattern of data analysis. These technologies enable analysts to work with a large flow of high-performing data – often information is created and collected in real-time, a clear example being e-commerce companies that capture every act the customer performs on their website, from the time spent on each page to the purchase patterns (Ylijoki and Porras, 2016; Gandomi and Haider, 2015).

Velocity refers to the speed at which data is received and stored by the system, a simple example would be an ad served on Google showing the number of clicks in real-time. As machine processing capabilities advance, technology tends to offer faster and more efficient storage of information. In a scenario with the continuous flow of large amounts of data, the tool needs to have high analysis performance so that patterns can be found quickly. Cloud computing is one of big data's main "allies." By running such systems in the cloud, analysts gain greater operational scalability at lower cost. Thus, if the flow of information increases, it is possible to scale resources, preventing the new demand impacting the speed of execution of analysis routines (Ylijoki and Porras, 2016; Gandomi and Haider, 2015).

Variety is another aspect of working with varied data streams, as the information can come from diverse devices, social networks, mobile devices and even offline media such as market research and financial transaction data tables, being the characteristic related to the types of data collected, where depending on it each information can have a distinct relevance during the analysis. With this focus, a big data solution must be able to handle different types of content; where it is the result of poor programming, the computational cost for analysts to perform their work will be high, while still generating a low reliability of the insights obtained (Ylijoki and Porras, 2016; Gandomi and Haider, 2015).

Thus, data have particular characteristics and, therefore, are classified into two types. One is unstructured data, which is used to define data that are not collected from a database or those that do not have a defined structure. Unstructured records should be cataloged before analysis, since they have a higher number of noises (information not relevant to the analyst). This is usually the case for text messages, emails, Word documents, PowerPoint presentations and even media files (audio and video). Unlike structured data, which is when collected data is already organised in a database or similar solution and is easier to evaluate by big data tools, since the number of routines that must be performed for the content rating is lower (Ylijoki and Porras, 2016; Gandomi and Haider, 2015).

Veracity is related to measuring the level of reliability of the data, where the accuracy of the information is important so as not to make misleading analyses about a given situation. To ensure that data analysis is able to meet business needs, it is crucial that the company can work with reliable data sets; often unstructured data records can lead to scenarios where the level of noise is high, impacting the quality of the analyst's work (Ylijoki and Porras, 2016; Gandomi and Haider, 2015).

Thus, big data solutions should look for data from trusted sources, and give you the ability to filter out which content is relevant to the business, and eliminate unreliable or unrelated content. Security should still be taken into account and is an essential aspect, especially due to the new Data Protection Regulations, where maintaining the integrity of information regarding storage, sharing and analysis is essential, avoiding leaks and misuse provided by national and international law (Ylijoki and Porras, 2016; Gandomi and Haider, 2015).

The solution must be able to add value to processes and make services more competitive, regarding the possible uses of information, identifying trends and patterns that enable managers to make decisions with confidence and better direct the strategy to win customers and/or more markets. Even for the operational aspect, you can evaluate internal routines and use of corporate tools to track bottlenecks and make process management more efficient. Value is directly related to collecting a large amount of data, and in the meantime storing things that will not be useful for improving the sale of any product or service (Ylijoki and Porras, 2016; Gandomi and Haider, 2015).

Thus, big data can be summed up to be used to define a large set of IT tools that enable the capture, analysis and cataloging of records in real time, where this information can be sourced from different internal and external sources such as data entry. customers, market analytics, social networking, electronic devices, internal processes or even offline media surveys. The advantage of this use lies in centralising the collection and analysis of this large set of records. From this, statistics and processing techniques are left to machines, allowing analysts to quickly identify patterns and predict trends more accurately (Monino and Sedkaoui, 2016).

One mistake in thinking about the importance of big data is to consider only the amount of data available, since what you need to consider is what do with it. Big data is important to help companies analyze their data and use it to identify new opportunities, obviously varying according to the business model and the activity performed as well as the goals and objectives set for the organisation. But there are benefits in common for every type of business, including time savings, cost savings, offer optimisation, new product delivery, higher profits, happier customers and more efficient decisions. Data are considered to be the new oil and those who know how to deal with them gain great advantage in the market. That's why big data is receiving increasing investments from companies of all sizes, and the professionals dealing with them are in intense competition around the world (Porter and Heppelmann, 2015).

8.3 Cloud Computing

Cloud computing is a concept that refers to a technology that allows access to programs, files and services through the Internet, without the need to install programs or data storage, hence the allusion to "cloud." This computation is a kind of cloud that gives individuals and businesses the ability to pool well-maintained, secure, easily accessible and on-demand computing resources. Cloud computing is the provisioning of virtualised IT resources over the Internet; it is on-demand computing as a service, with pay-as-you-go, through a cloud service platform. As a result, services can be accessed remotely from anywhere in the world, at any time (Rittinghouse and Ransome, 2017).

The "cloud" is not a place, but a method of managing (Rittinghouse and Ransome, 2017) IT resources by replacing local machines and private data centers with a virtual infrastructure, where users access virtual computing, networking and storage resources made available online by a remote provider. These features can be provided instantly, which is particularly useful for companies that need to scale up or down when responding to fluctuations in demand. Cloud computing is a technology that enables the distribution of your computing services and online access to them without the need to install programs, precisely because you do not need to install programs or store data (Sen, 2015).

There are three different ways to deploy cloud services: public cloud, private cloud or hybrid cloud. A public cloud belongs to a third-party cloud service provider by which it is administered. This provider provides cloud computing capabilities such as servers and web storage. With a public cloud, all hardware, software and support infrastructures used are owned and managed by the cloud provider hired by the contracting organisation, which can access these services and manage your account using only one web browser. The main benefits of the public cloud are on-demand scalability and pay-as-you-go pricing (Sen, 2015).

Private cloud refers to cloud computing resources used exclusively by a single company and may be physically located in the company's local data center. That is, a private cloud is one in which the cloud computing services and infrastructure used by the company are maintained on a private network, supposing that you have the option of hiring outsourced cloud computing service providers to host your private cloud. This type of cloud works behind a firewall on a company's intranet and is hosted in a data center dedicated to such an organisation. Private cloud infrastructure can be configured and managed to meet the individual needs of an enterprise (Sen, 2015; Wang et al., 2016).

Hybrid cloud is the combination of a public and a private cloud, which are linked by a technology that allows data and application sharing between them. As its name suggests, the hybrid cloud model allows companies to use both public and private cloud solutions. With a hybrid cloud, organisations can leverage the strengths of each cloud model to enable flexibility and scalability while protecting sensitive data and operations. This shared data and applications can move between private and public clouds, giving your business greater flexibility and more deployment options (Sen, 2015; Patel et al., 2016).

It is also possible to classify cloud storage according to the technology used. In all-flash, data is stored in flash memory with very high read and write speeds. As standard, the most common model used is flash memory where "hot" data is stored, and discs where "cold" data is stored. A more business-oriented model, is the enterprise model of performance disks, where speed is not a big attraction, but fault tolerance is an advantage. There is still backup-specific cloud storage that utilises volumetric disks and features low performance, affordability and plenty of available space (Rittinghouse and Ransome, 2017).

Most cloud computing services fall into four broad categories: infrastructure as a service (IaaS), platform as a service (PaaS), serverless and software as a service (SaaS), as well as what can be called the pillars of cloud computing. because one is theoretically based on the other. IaaS is the most basic category of cloud computing, with it leasing the IT infrastructure of a cloud service provider, paying only for its use and hiring services involving the acquisition of servers and virtual machines, storage (VMs), networks and operating systems. IaaS provides users with access to basic infrastructure aspects such as server space, data storage and networking that can be provisioned through an API, which is the model that most closely replicates the functionality of a traditional data center in a hosted environment (Madni et al., 2016).

PaaS refers to cloud computing services that provide an on-demand environment for software application development, testing, provisioning and management, so developers do not have to deal directly with the infrastructure layer when deploying or upgrading applications. It is designed to make it easier for developers to build mobile or web applications, making it much faster, eliminating concerns about configuring or managing the underlying infrastructure of servers, storage, networking and databases needed for development (Pahl, 2015).

SaaS is a method for distributing software applications over the Internet on demand, and typically subscription-based applications are designed for end users. With all the provisioning and infrastructure development going on behind the scenes, users can connect the application over the Internet, usually with a web browser on your phone, tablet or PC. Cloud providers host and manage the software application and underlying infrastructure, as well as perform maintenance such as software updates and security patching. For business applications, SaaS offers a wide range of cloud functionality, from word processing and spreadsheet programs, photo editing packages to video hosting platforms (Dašic et al., 2016).

Serverless computing, like PaaS, focuses on building applications without wasting time on the ongoing management of the servers and infrastructure needed to do so. Where the cloud provider handles all server configuration, capacity planning and management, serverless architectures are still highly scalable and event-driven, using resources only when a function or event that triggers such a need occurs (McGrath and Brenner, 2017).

Data flexibility can be realised more cost-effectively, and on a large scale, with the ability to transfer to a cloud storage system. Analyze and unify a company's data across teams, divisions and locations, using cloud services for machine learning and artificial intelligence to uncover insights and make better-informed decisions. Cloud computing enables the use of intelligent models (inserting and generating intelligence) to help engage customers and provide important insights based on captured data (Rittinghouse and Ransome, 2017).

However, cloud computing also has relative disadvantages when it comes to data security, where many people are suspicious and uncomfortable with important information in a virtual environment. And to avoid this kind of problem, computer supply companies invest heavily in security and "cloud antivirus." Security is one of the biggest concerns for companies looking to move all or part of their IT operations to the cloud, so in some segments, data security compliance regulations require some applications to stay in private data centers, which brings about the need for private or hybrid cloud models. Another important factor is having a stable and fast connection to get the most out of the technology, as the core of the technology needs it to access remote servers, especially when it comes to streaming and gaming. Because servers are far away, unstable or slow Internet speeds are detrimental to the full use of technology (Kumar and Charu, 2015).

8.4 Big Data and Cloud Computing

The cloud services market has grown so much that now a remarkable share of total IT spend helps create a new generation of cloud-born startups and providers that, in addition to disruptive, cloud computing strategies are the basis for remaining relevant in a fast-paced world. IT spending has constantly shifted from traditional offerings to cloud services. Many markets will be indirectly affected, such as agribusinesses, medicine, logistics and so on, since big data storage, with the later use of data analytics to identify relevant decision-making information, is a process that has become increasingly common and its results demonstrate that it is possible to add value from the cloud data and services architecture. As organisations pursue a new IT architecture and operational philosophy, they are gearing up for new opportunities in digital business, including next-generation IT solutions such as the Internet of Things (IoT) (Assunção et al., 2015).

The trend of more people doing more with their data will continue quickly, so the spread of self-service data analytics, along with widespread cloud adoption, is causing many changes that businesses need to take advantage of. And with the continued presence of big data, the use of softwares that represent complex statistical and mathematical models to analyze large amounts of data helps in bringing information such as prediction of market behavior, suggestions for process improvements and understanding more in-depth scenarios such as economic outlook and consumer needs. Big data refers to a tremendous amount of data and the cloud, Internet-accessible digital storage, and processing framework, which together make the services and internal processes of an enterprise scalable (Yang et al., 2017)

Extra processing is the main reason that drives a company to analyze data using the cloud, since big data requires high computing capacity to be interpreted. The cloud offers this analytics capability. It provides access to resources such as servers and networks in an organised, practical and cost-effective manner, favoring the storage and processing of large amounts of data, especially when demand is increasing over time, i.e., big data. Cloud computing services custom-tailor the exact amount of resources as needed, providing extra power quickly, in a much shorter time than would be required to upgrade and modify physical servers where technology allows upgrades quickly and with intelligent use of resources, i.e., scalability (Assunção et al., 2015).

One of the key factors that makes the cloud more secure from physical devices for working with big data is the timeliness of the information, i.e., it is available indefinitely. This ensures traditional hard drives (on-premise) are free of major flaws, while cloud services rely on specialised support from vendors, making data longevity virtually endless. Cloud computing has at least five features that are essential for effective big data application, being: **on-demand structure**, where hiring a cloud infrastructure is tailored to business needs; **ubiquity**, related to the storage of data in the cloud, which can be accessed and collected from anywhere with Internet access and through the most diverse devices; **scalability**, where the amount of information stored grows, as well as the demand for processing power (happening continuously and rapidly for big data analytics), and more physical or virtual resources used in the cloud can be easily allocated; **elasticity**, which is related to computational resources that are changed transparently to service users without interrupting the use or processing of information; and **monitoring and measurement**, through a web platform or a mobile application, being able to monitor the amount of resources used, the equipment performance and the amount used in relation to the contracted capacity (Lv et al., 2017).

In this context, every company has data that can help them make more profit. The benefits of combining cloud computing with big data can be applied to streamline analytics, strategise and better understand the industry in which your company operates with big data in cloud for fast analytics with new memory and computing options. This gap between demand and viability can be narrowed by connecting big data with elements of cloud computing, since the cloud is a necessary prerequisite for effectively using big data, where the necessary IT resources are always there, available on demand and scalable, if necessary, to the minimum investment. The alliance between cloud and big data brings new integrated methods of analysis, the result of which directly involves business decisions and can also generate new processes (Lu, 2017).

8.5 Discrete Event, Communication Channel and Modulation

The discrete event technique is robust in its abstraction, which can be applied in the broad context of a communication system, as well as in the transmission of data in a channel, characterised by occurrence events responsible for changes in the state of the system in which they act, producing state changes at random intervals of time, used to model and represent the system as a sequence of operations through entities, transactions, of specific types such as bits, these being derived from the results of actions taken in a system, which can be classified as generating data and therefore the information, just as it has been used and is used to model concepts with a high level of abstraction, that is, the exchange of emails on a server to transmission, data packets, etc. The entities are discrete in discrete event simulation, an instantaneous or discrete incident changing the state variable, or an occurrence coming from event, being the results of actions that occur in the system, usually producing state changes at random time intervals, being defined as discrete items of interest in a discrete event simulation, where its meaning depends on what is being modeled and the type of system used (Dagkakis and Heavey, 2017).

Faithful modeling of a communication channel is the medium that provides the physical connection between transmitters and receivers, either wired or a logical connection in a multiplexed medium, considering noise, which changes some of the characteristics of the signal transmitted by another external signal to the transmission system or generated by the transmission systems themselves. This is the model widely used for a large set of Additive White Gaussian Noise (AWGN) physical channels, with the characteristic of modeling and statistically representing these types of noise entering the transmitted signals. Still, considering the fading of a channel can be large-scale fading, where there is attenuation of the average power or loss in the signal path due to movement of the receiver in large areas, Rician is advised because the dominant component of the received signal is stationary, i.e., when there is a channel normally encountered with a predominant line of sight or the dominant route, considering that multiple replicas of the transmitted signal may arrive with different attenuations and generate different delays in the receiver (Sharma and Singh 2016)

Modulation is the technique where the characteristics of the signal (carrier) are modulated, where they are modified for the purpose of transmitting the information, making the combined changes of frequency, amplitude or phase. This modulated carrier wave travels over a communications channel carrying all information. Thus, a specific type is Differential Quadrature Phase Shift Keying (DQPSK) format, a particular form of Quadrature Phase Shift Keying (QPSK) modulation, rather than a symbol corresponding to a pure phase parameter; this symbol represents a phase variation, determined by 4 possible states 0, π, $+ \pi/2$, $-\pi/2$, where constellation rotated by

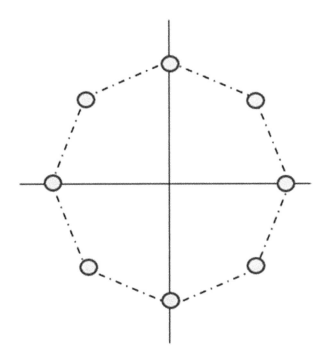

FIGURE 8.1
Theoretical DQPSK constellation.

45 ° from the previous point, referring that each symbol represents two bits of information and shifted to about $\pi/4$ or $\pi/2$, totaling 8 status positions, still taking into account withstanding the channel fading, being used by most of the cellular towers for transmission of the data and long-distance communication. Figure 8.1 shows the DBPSK constellation diagram (Bhat and Singh, 2017).

8.6 Methodology

Was used the Simulink tool, from Matlab (2014a), chosen because it is already consolidated in the scientific medium, with an IDE (Integrated Development Environment) already tested and validated, still using the Communications System library, designed to design, simulate and analyze systems, able to model dynamic communication systems; the DSP System, capable of designing and simulating systems with signal processing; Simulink's own library, for block diagram environment for multi-domain simulation; and SimEvents, classified as a discrete event simulation mechanism and components to develop systems models oriented to specific events.

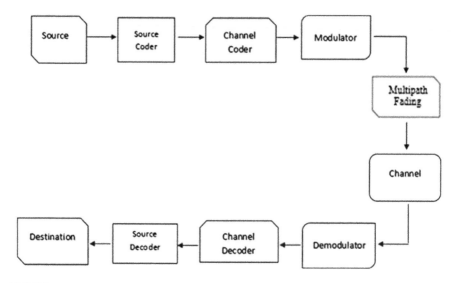

FIGURE 8.2
Traditional model of a telecommunication system.

Previous research (França et al., 2020; R. I. Padilha, 2018; R. F. Padilha 2018) proved that bit 0 treatment is feasible, showing significant results in the compression of the information transmitted on a channel, likewise in a complementary manner in the reduction of processing time. So, the proposed method is based on the development of an AWGN channel particularised by the implementation of the discrete-event technique in the bit generation step.

In the proposed model (see Figure 8.2), the signals corresponding to bits 0 and 1 will be generated and modulated with the advanced modulation format DQPSK, passing through a multipath Rician fading channel with Jakes model with Doppler shift defined at 0.01 Hz and also inserted a block with a math function 1/u that is required to track the channel time variability where the receiver implementation ordinarily incorporates an automatic gain control (AGC) and will proceed to an AWGN channel according to the parameters shown in Table 8.1, signal will then be demodulated.

TABLE 8.1

Parameters Channel Models DQPSK Rician

AWGN DQPSK	
Sample Time	1 sec
Simulation time	10000 sec
Eb/N0	0 a 12 dB
Symbol period	1 sec
Input signal power	1 watt
seed in the generator	37
seed on the channel	67

FIGURE 8.3
Proposed bit precoding.

The modeling, according to the proposal implemented with discrete events is similar to the one presented previously, was implemented through the discrete event methodology, understood as the discrete event methodology in the step of generating signal bits (information) to make it more appropriate for a specific application. The event-based signal is a signal susceptible to treatment by the SimEvents library, and posteriorly passed by conversion to the specific format required for manipulation by the Simulink library, where both time-based signals and event-based signals were in the time domain, which was generated as a discrete entity and followed the parameters as presented in Table 8.1. Then, Entity Sink represents the end of the modeling of discrete events by the SimEvents library.

Entity Sink will be located by SimEvents, where the event-based signal conversion will be performed for a time-based signal, being converted to a specific type that followed the desired output data parameter, an integer, the bit. By means of the Real-World Value (RWV) function, the current value of the input signal was preserved. Then a rounding was performed with the floor function. This function is responsible for rounding the values to the nearest smallest integer. Zero-Order Hold (ZOH) is used for discrete samples at regular intervals, describing the effect of converting a signal to the time domain, causing its reconstruction and maintaining each sample value for a specific time interval. The treatment logic on bits 1 and 0 is shown in Figure 8.3. Subsequently, the signal is modulated with the advanced modulation format DQPSK and is inserted into the AWGN channel, and then demodulated for the purposes of calculating the BER of the signal, as shown in Figure 8.4.

The models are shown in Figures 8.2 and 8.4 and run with 10000 seconds of simulation, respecting the configuration defined according to Table 8.1.

The constellation has as function to analyze both signals transmitted by the models; this validation methodology has as function to affirm that the proposal will not modify the amount of bits transmitted by the signal, since both signals transmitted in the traditional channel, and in the channel containing the proposal of this study, will be of the same size.

8.7 Results and Discussion

In this section AWGN transmission channel with DQPSK modulation will be presented. Figure 8.5 shows the traditional method (left) and the proposed

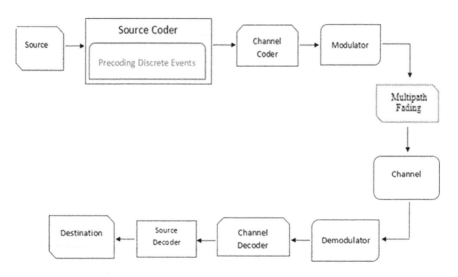

FIGURE 8.4
Model of a telecommunication system with the proposal.

method (right), showing the signal transmission flow (corresponding to bits 0 and 1), generated and then modulated, passing through the channel. Figure 8.6 displays the constellations for 19 dB for the proposed (left) and the traditional methods (right).

The models were investigated from the perspective of memory consumption evaluation. Because the first simulation in each command is analyzed, the construction of the model in a virtual environment is performed from scratch; where all the variables of the model are allocated, the memory of the operating system in which the MATLAB is running is reserved for execution, according to the evaluation parameters are, in fact, real.

For memory calculation, the "sldiagnostics" function was used, where the "TotalMemory" variable received the sum of all the memory consumption processes used in the model, by the "ProcessMemUsage" parameter, counting the amount of memory used in each process, throughout the simulation, returning the total in MB (megabyte). The physical structure used for the simulation was the Intel Core i3 processor computer with hardware configuration, with two-color processing, Intel Hyper-Threading Technology and 4GB RAM, and this machine relates to the dynamics of real-world efficiency and applicability. The experiments were carried out through three simulations of each model developed in order to develop the analysis of this chapter, as shown in Figure 8.7. The respective values of memory consumption are found in Table 8.2.

Utilising the capabilities offered by cloud computing facilitates the analysis of large data volumes and ensures competitive advantages for companies. Big data is a technique that needs high processing and storage potential to efficiently transform large data flow in useful information for

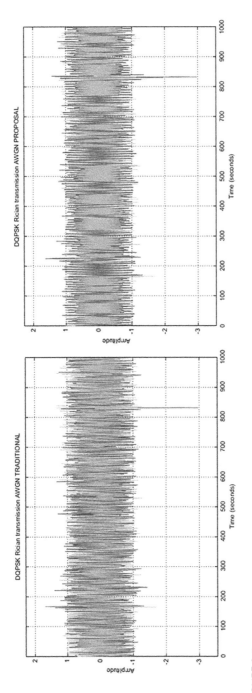

FIGURE 8.5
Transmission flow DQPSK Rician.

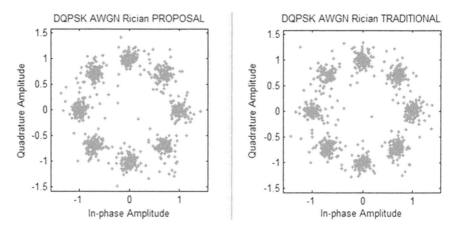

FIGURE 8.6
Simulated DQPSK Rician constellations.

Memory Consumption

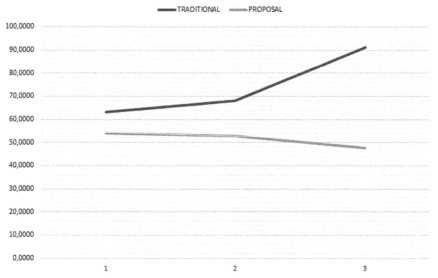

FIGURE 8.7
First simulations (memory) model DQPSK Rician.

the company. Because of this factor, the cloud environment and its resources is a great ally for the development of these strategies and analysis. Together they provide a solution that is scalable and adaptable for large data sets and business analytics, where the advantage of analytics becomes the great benefit. Related to much of the information that seeks and generates these days is on the network, where things of interest are constantly searched in

TABLE 8.2

Memory Consumption

	Traditional	Proposal
Simulations		
1	63,2109	54,0820
2	67,9648	52,7422
3	91,0820	47,6094

search engines, doing different tasks through electronic devices, buying products online and many other activities that somehow involve the use of the internet.

All this data, when in the hands of a company, becomes valuable insight into the behavior of its target audience. In general, companies use big data to become more competitive, because it is expected to be less erroneous, which is essential for success. A cloud computing system is a prerequisite for working well with a large amount of data, as it involves collecting, storing and sharing a huge amount of information, where all that information can bring valuable insights.

In this focus, the proposal presented brings a new approach to signal transmission, since it is performed in the discrete domain with the implementation of discrete entities in the bit generation step. This process is the result of a methodology used in a lower level of application, which acts on the physical layer, and is usually used in the transport layer to increase the capacity of information transmission for communication systems.

Generating smoother transmission and communication of the system as a whole, since slowness is related to the speed of communication and the cloud technology implemented in the system structure used by users, can often generate inconvenience to the user, due to crashes, and sometimes even loss of data, directly impacting the improvement of cloud environments with big data. Just as reducing the consumption of computational resources, such as memory, generate greater reliability and can be seen as a great ally, meeting the needs of a technological world found today. Due to the proposal to demonstrate a lower reaching up to 91.31% memory consumption, compared to the traditional methodologies, which do not require the use of discrete events in data transmission.

8.8 Future Research Directions

Future research is related to the development and implementation of the proposal presented in other digital communication systems considering Multiple Input and Multiple Output (MIMO) as well as modern systems such as Orthogonal Frequency Division Multiplexing (OFDM), including simulations on a wider variety of hardware. The proposal with modern applicable bias might be in architectures like Intel i5 and i7 Processors, or other multi-core processors.

8.9 Conclusion

More organisations are storing, processing and extracting value from data of all types and sizes, where digital systems must support large volumes of structured and unstructured data, where this need will continue to grow. Thus, there will be market demand for platforms, solutions and new methodologies that help data administrators govern and protect big data and enable users to analyze this data, where these systems will mature to operate seamlessly with cloud technology. Cloud computing and big data are two concepts that work interdependently and are transforming the way businesses perform their functions. Cloud computing organises all of the data storage, making it accessible and simplifying the routine, so making it a facilitator for the use of big data, which provides the necessary capacity to handle such large amounts of data. The two concepts start from dematerialisation logic, where since they do not exist in a physical space, they are accessible to anyone who has access to the Internet. Thus, there are numerous benefits to those who make use of these two technologies, which are different from each other but are interrelated.

This chapter has shown that the developed method has great potential in the improvement of communication services potential, useful for both cloud and big data, since the difference is that there is no use of discrete events applied in the physical layer (the bit itself) of transmission. Transmission medium has shown better computational performance in terms of memory utilisation by compressing the information to 91.31%. Proportionally linked cloud computing services and big data, due to the fact that the flow of data will consume fewer resources, can improve interactions between the technologies, presenting more efficient and productive processes, aiming at efficiency in the transmission of data and information.

References

Assunção, M. D., Calheiros, R. N., Bianchi, S., Netto, M. A. and Buyya, R., 2015, 'Big data computing and clouds: trends and future directions', *Journal of Parallel and Distributed Computing*, pp. 3–15.

Bhat, S. A. and Singh, A., 2017, 'Review on effective image communication models', *Australian Journal of Basic and Applied Sciences*, pp. 65–79.

Bihl, T. J., Young II, W. A. and Weckman, G. R., 2016, 'Defining, understanding, and addressing big data', *International Journal of Business Analytics (IJBAN)*, pp. 1–32.

Dagkakis, G. and Heavey, C., 2017, 'A review of open source discrete event simulation software for operations research', *Journal of Simulation*, pp. 193–206.

Dašic, P., Dašic, J. and Crvenkovic, B., 2016, 'Service models for cloud computing: search as a service (SaaS)', *International Journal of Engineering and Technology*, pp. 2366–2273.

Erevelles, S., Fukawa, N. and Swayne, L., 2016, 'Big data consumer analytics and the transformation of marketing', *Journal of Business Research*, pp. 97–904.

França, R. P., Iano, Y., Monteiro, A. C. B. and Arthur, R., 2020, 'Improvement of the transmission of information for ICT techniques through CBEDE methodology', in B. Chintan, P. S. Sajja and S. Liyanage (eds.), *Utilizing educational data mining techniques for improved learning: emerging research and opportunities*, IGI Global, pp. 13–34.

Gandomi, A. and Haider, M., 2015, 'Beyond the hype: big data concepts, methods, and analytics', *International Journal of Information Management*, pp. 137–144.

Grover, V., Chiang, R. H., Liang, T. P. and Zhang, D., 2018, 'Creating strategic business value from big data analytics: a research framework', *Journal of Management Information Systems*, pp. 388–423.

Kumar, R. and Charu, S., 2015, 'Comparison between cloud computing, grid computing, cluster computing and virtualization', *International Journal of Modern Computer Science and Applications*, pp. 42–47.

Loebbecke, C. and Picot, A., 2015, 'Reflections on societal and business model transformation arising from digitization and big data analytics: a research agenda', *The Journal of Strategic Information Systems*, pp. 149–157.

Lu, Y., 2017, 'Industry 4.0: a survey on technologies, applications and open research issues', *Journal of Industrial Information Integration*, pp. 6–10.

Lv, Z., Song, H., Basanta-Val, P., Steed, A. and Jo, M., 2017, 'Next-generation big data analytics: state of the art, challenges, and future research topics', *IEEE Transactions on Industrial Informatics*, pp. 1891–1899.

Madni, S. H. H., Latiff, M. S. A. and Coulibaly, Y., 2016, 'Resource scheduling for infrastructure as a service (IaaS) in cloud computing: challenges and opportunities', *Journal of Network and Computer Applications*, pp. 173–200.

McGrath, G. and Brenner, P., 2017, 'Serverless computing: design, implementation, and performance', in *2017 IEEE 37th international conference on distributed computing systems workshops (ICDCSW)*, IEEE, pp. 405–410.

Monino, J. L. and Sedkaoui, S., 2016, *Big data, open data and data development*, Vol. 3, John Wiley & Sons.

Oussous, A., Benjelloun, F. Z., Lahcen, A. A. and Belfkih, S., 2018, 'Big data technologies: a survey', *Journal of King Saud University-Computer and Information Sciences*, pp. 431–448.

Padilha, R., Iano, Y., Monteiro, A. C. B., Arthur, R. and Estrela, V. V., 2018, 'Betterment proposal to multipath fading channels potential to MIMO systems', in Brazilian technology symposium, Springer, pp. 115–130.

Padilha, R. F., 2018, 'Proposta de um método complementar de compressão de dados por meio da metodologia de eventos discretos aplicada em um baixo nível de abstração= Proposal of a complementary method of data compression by discrete event methodology applied at a low level of abstraction', *UNIICAMP*.

Pahl, C., 2015, 'Containerization and the paas cloud', *IEEE Cloud Computing*, pp. 24–31.

Patel, A., Patel, M. and Singh, P. K., 2016, 'Study of security in the hybrid cloud', *International Journal for Research in Advanced Computer Science and Engineering*, pp. 01–05.

Porter, M. E. and Heppelmann, J. E., 2015, 'How smart, connected products are transforming companies', *Harvard Business Review*, pp. 96–114.

Rittinghouse, J. W. and Ransome, J. F., 2017, *Cloud computing: implementation, management, and security*, CRC Press, Boca Raton, FL.

Sen, J., 2015, 'Security and privacy issues in cloud computing', in Management Association (ed.), *Cloud technology: concepts, methodologies, tools, and applications*, IGI Global, Hershey, PA, pp. 1585–1630.

Sharma, S. and Singh, H., 2016, 'Comparison of different digital modulation techniques in LTE system using OFDM AWGN channel: a review', *International Journal of Computer Applications*, pp. 1–4.

Wang, H., He, D. and Tang, S., 2016, 'Identity-based proxy-oriented data uploading and remote data integrity checking in public cloud', *IEEE Transactions on Information Forensics and Security*, pp. 1165–1176.

Yang, C., Huang, Q., Li, Z., Liu, K. and Hu, F., 2017, 'Big data and cloud computing: innovation opportunities and challenges', *International Journal of Digital Earth*, pp. 13–53.

Ylijoki, O. and Porras, J., 2016, 'Perspectives to definition of big data: a mapping study and discussion', *Journal of Innovation Management*, pp. 69–91.

Zomaya, A. Y. and Sakr, S., 2017, *Handbook of big data technologies*, Springer, Berlin.

9

Heterogeneous Data Fusion for Healthcare Monitoring: A Survey

Shrida Kalamkar and Geetha Mary A

CONTENTS

9.1 Introduction

With the emergence of the Internet of Things, where all kinds of devices and systems are connected with each other, an enormous amount of data is continuously being generated in all walks of life. A significant portion of such data is being captured, stored, aggregated and analyzed systematically without losing its "4Vs" characteristics, i.e., (volume, velocity, variety and veracity). This big data Internet of things revolution is under way in healthcare (Lee et al., 2008). The term "Internet of Things" was firstly coined by Kevin Ashton in a presentation in 1998 (Ashton, 2009). He also mentioned that "The IoT has the potential to change the world, just as the Internet did. Maybe even more so" (Sundmaeker, Guillemin and Friess, 2010). Then, MIT presented their IoT vision in 1999. Later, the IoT was formally introduced by the International Telecommunication Union (ITU) by an ITU Internet report in 2005 (ITU, 2005). The goal of the Internet of Things is to enable things to be connected anytime, anyplace, with anything and anyone ideally using any path/network and any service (Vermesan and Peter Friess, 2013).

In recent years, healthcare applications such as home-based care, disaster relief management, medical facility management and sports health management have attracted considerable interest (Paganelli, Spinicci and Giuli, 2008; Luprano et al., 2007). Big data and pervasive technologies are increasingly used for biomedical and healthcare informatics research. Large amounts of biological and clinical data have been generated and collected at an unprecedented speed and scale. There are several technologies that allow large amounts of data to be handled. But despite the evolution of big data processing technologies such as Hadoop and Apache Storm and scalable infrastructure such as virtualised clouds, there still remains a significant gap with regard to heterogeneous data collection and real-time analysis based on defined quality of service (QoS) constraints (Shah et al., 2016). Given the increase in volume, velocity and variety of healthcare data sensors, special techniques and technologies for analysis and inferencing are required.

This chapter is intended to first provide an overview of data fusion techniques. Later the importance of data fusion in healthcare application is highlighted. Section 9.4 discusses the survey of various healthcare sensor data fusion efforts and its importance for the IoT. Later issues in handling healthcare data are discussed. Section 9.5 highlights the data fusion framework metrics from a healthcare perspective. Section 9.6 discuss the applications of data fusion in healthcare. The final section concludes the survey by highlighting the survey results and research gaps.

9.2 Sensor Data Fusion

Sensor data fusion is an essential and integral part of the Internet of Things (IoT). Sensors are used in variety of applications, such as climate monitoring, smart mobile devices, healthcare, automotive systems, industrial control, traffic control. Data in IoT is dynamic and heterogeneous, which leads to inadequacy of simple single-source analysis methods. Data fusion integrates multiple data and knowledge into a consistent, accurate and useful representation which makes data fusion to provide high-quality information for a reliable decision support. Data fusion also leads to an increase in the accuracy of information generated from multiple sources by reducing the uncertainty of data. The fusion of complementary information generated from multiple sources can provide more accurate information instead of single-source information. Confidence can be increased when multiple independent measurements are made on the same event dataset. Therefore, the result is more reliable. In the world of the IoT, as the size of data increases, handling these large volumes of streaming and historical data, which can vary from structured to unstructured and numerical to microblog data streams, is challenging because its volume is heterogeneous and highly dynamic. Hence, techniques and methodology for understanding and resolving issues about data fusion in the IoT needs to be investigated. The data sources for a fusion process are not specified to originate from identical sensors. Therefore, the data fusion process is categorised into two types, as shown in Figure 9.1.

The direct fusion process is the fusion of sensor data from a set of heterogeneous or homogeneous sensors, soft sensors, and history values of sensor data.

The indirect fusion process uses information sources like a priori knowledge about the environment and human input.

Sensor fusion can be divided in three categories depending on the extent of the data processing that occurs in each sensor, the data products produced by the individual sensors and the location of the fusion processes (A.Klein, 2004). Figure 9.2 gives an overview of sensor fusion categorisation.

1. Sensor level (low level): In this level of fusion, raw data are acquired from diverse sensors.
2. Feature level (intermediate level): Here features are extracted from each sensor or sensor channel and combined into a composite feature, representing an object. These features are combined to form a composite vector that is input to a neural network.
3. Decision level (high level): Decision-level fusion is associated with sensor-level fusion. The results of the initial object detection and classification by the individual sensors are input to a fusion algorithm.

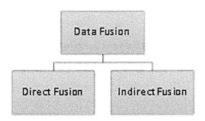

FIGURE 9.1
Fusion process.

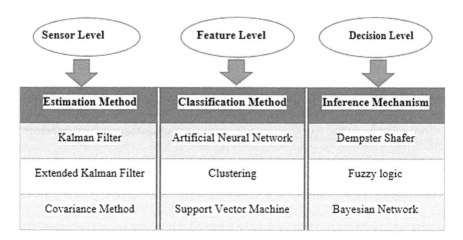

FIGURE 9.2
Sensor fusion categorisation.

> Final classification occurs in the fusion processor using an algorithm
> that combines the detection, classification and position attributes of
> the objects located by each sensor.

The design and real-time implementation of fusion systems is an extremely
complex task. Requirements analysis, sensor selection, architecture selection, algorithm selection, software implementation, testing and evaluation
are some of the issues related to the implementation of data fusion systems.
Hence there is a need for generic interface to ingest data from multiple interfaces and IoT systems, store it in a scalable manner and analyze it in real
time to provide an efficient, scalable and reliable solution for IoT applications. A number of techniques, such as Bayesian method, Dempster–Shafer's
method, Fuzzy logic, artificial neural networks, Kalman filter and Extended
Kalman Filter can be used in sensor data fusion (https://en.wikipedia.org/
wiki/Data_fusion, no date).

In Table 9.1 comparative analysis of sensor fusion techniques is presented.
These algorithms are the most frequently used algorithms for sensor fusion.
Detailed study of each sensor fusion algorithm is presented along with
advantages, limitations and most probable application areas of the algorithm.

9.2.1 Sensor Data Fusion in the Healthcare Environment

Healthcare data fusion in the IoT environment is one of the most important challenges that needs to be addressed to develop innovative services.
Heterogeneous medical data have been generated in various healthcare
organisations, including payers, medicine providers, pharmaceutical

TABLE 9.1

Comparative Analysis of Sensor Fusion Algorithms

Algorithm	Advantages	Limitations	Application Areas
Kalman Filter (KF) (Thomas, 2010; Sasiadek and Hartana, 2000)	• Easy to implement • Used in Linear systems • Efficient computation	• It assumes that both the system and observation models equations are both linear, which is not realistic in many real life situations • It assumes that the state belief is Gaussian distributed. • Accuracy issues	Guidance, navigation and control of vehicles Computer vision applications
Extended Kalman Filter (EKF) (Ant, 2016)	• Used in non-linear systems and non-Gaussian uncertainty problems • More efficient than KF	• Inconsistent results • High computational complexity • Fails if the functions are highly nonlinear	Non-linear problems like tracking of the target etc.
Covariance Methods (Thomas, 2010)	• Used in decentralized systems • Easy to implement • Accurate results	• Conflicting behavior in certain cases	Genetics and molecular biology, financial economics, meteorological and oceanographic data assimilation, feature extraction
Dempster–Shafer (Wu et al., 2002)	• Combines different sources of evidence • No prior information needed • Deals with statistical problems • Less available sensors can draw good evidences • Safe and reliable	• Computational complexity with increase in number of sensors • Lack of well-established decision theory • Belief and probability may differ • D–S technique depends mainly on the quality of basic detectors	Aircraft maintenance

(Continued)

TABLE 9.1 (CONTINUED)

Comparative Analysis of Sensor Fusion Algorithms

Algorithm	Advantages	Limitations	Application Areas
Fuzzy Logic (Gibson, Hall and Stover, 1994)	• High robustness and good utility • Decision-based method • Fast response to rapid changes • Contextual representation • Requires a minimum degree of expertise	• Complex computation • Confusing results • Does not deal with novel situation since it is related to a predefined knowledge repository and databases	Machine diagnosis and conditioning for complex production machines and process engineering detection decisions, fuzzy correlation
Bayesian network (Junghans and Jentschel, 2007)	• Suitable for multi-source information • easily extendable for higher number of heterogeneous sensors • well-established method in decision-level fusion	• High memory requirement • Performs well in well-established environment • Time complexity is higher	Traffic Management, Healthcare, Semantic Search, Image Processing and information retrieval Biomonitoring
Artificial Neural Network (ANN) (Fincher and Mix, 1990)	• Handles non-linear data • Efficient for heterogeneous sensor data • Good for high-level inference	• High complexity • Unreliable results in some cases • Slow context inference	Practical applications like fusion of multiple sensors in pattern recognition, robot navigation and military environment assessment/evaluation.
Clustering	• Can perform well under a high cluster tracking environment • Generates result of multitarget tracking using multi-sensor data association • Higher accuracy	• Computational overhead • Cluster head formation is a difficult process	IDS, Military applications
Support Vector Machine (SVM) (Banerjee and Das, 2012)	• Effective for heterogeneous sensor data • Better results for inconsistent sensor data • Good computation speed and memory	• Sensitive to noise • Text categorisation problems are linearly separable	Face detection, text categorization, classification of images, bioinformatics, protein fold and remote homology detection, handwriting recognition, geo and environmental sciences

information, prescription information, doctor's notes or clinical records produced day by day. Healthcare monitoring applications require the use of several body sensors to regularly measure the health parameters as well as environmental sensors to monitor the ambient parameters and transmit contextual information to the patient's healthcare network. The sensors collect patients' clinical data and the state of their surroundings, which are then transmitted to the appropriate healthcare centers and providers. The availability of multiple heterogeneous datasets presents new opportunities to big data analysts, because the knowledge that can be acquired from combined data cannot be acquired from any individual source alone. However, the biases that emerge in heterogeneous environments require new analytical tools. Surender Reddy Yerva et al. (Yerva, Hoyoung and Aberer, 2012) address the issue of fusing data in the cloud as various smart applications rely on mobile cloud computing. They have presented a data fusion approach that fuses social and sensor data in a travel recommendation system that facilitates the fusion process over massively streaming data. This proposed framework enables tweet messages to be analyzed to extract people's moods depending on day, weather and location.

Federico Castanedo (Castanedo, 2013) reviews the most popular methods and techniques for performing data/information fusion. These methods and algorithms are presented using three different categories: (i) data association, (ii) state estimation and (iii) decision fusion. The author states that to determine whether the application of data/information fusion methods is feasible, we must evaluate the computational cost of the process and the delay introduced in the communication. A centralised data fusion approach is theoretically optimal when there is no cost of transmission and there are sufficient computational resources.

Meisong Wang et al. (Wang et al., 2015) present a data fusion framework in a smart city based on ten metrics. They have evaluated the model based on performance, context awareness, cross domain portability, architecture, semantic integration, actuation management, implementation and fusion complexity. The authors have identified that context awareness in data fusion is still in its infancy.

Elias Bareinboim et J. Pearl (Bareinboim and Pearl, 2016) propose a general approach to data fusion problems, based on a syntactic transformation of the query of interest into a format derivable from the available information.

Yu Zheng (Zheng, 2015) explores the relationship and differences between different methods to understand proper data fusion methods to solve big data problems.

Furqan Alam et al. (Alam et al., 2017a) focus on mathematical methods and specific IoT environments (distributed, heterogeneous, nonlinear and object tracking environments) in data fusion. Adnan Akbar et al. (Akbar et al., 2018) propose a novel method for fusion of high-level events in real time based on complex event processing (CEP) with Bayesian networks (BNs) for large-scale IoT applications.

Hemanth Kumar et al. (Kumar and Pimparkar, 2018) explain a hierarchical approach for data fusion based on Fuzzy Kalman Filter for state estimation, and Dempster–Shafer method for decision fusion in IoT. Ibrar Yaqoob et al. (Yaqoob et al., 2016) discuss information fusion (IF) in social big data. The similarities and differences among big data technologies based on important parameters are analyzed. A discussion of the characteristics and current trends in social big data is provided.

Sandra Rodríguez-Valenzuela et al. (Rodríguez-Valenzuela et al., 2014) propose a distributed data fusion acquisition using a lightweight service composition model, which ensures the correctness of collaborations without a cyclic behaviour. This method allows working with data in a distributed, decentralised manner. Bakr and Lee (2017) discuss theories and methodologies of distributed multisensory data fusion with a specific focus on handling unknown correlation and data inconsistency. Jieping Ye et al. (Ye et al., 2008) have proposed a kernel method for integrating heterogeneous data for AD prediction. The kernel framework is extended for selecting features (biomarkers) from heterogeneous data sources. Experiments show the integration of multiple data sources leads to a considerable improvement in the prediction accuracy. Results also show that the multi-source feature selection algorithm identifies biomarkers (brain regions) that play more significant roles than others in AD diagnosis.

Furqan Alam et al. (Alam et al., 2017b) present a literature review on data fusion for IoT with a particular focus on mathematical methods (including probabilistic methods, artificial intelligence and theory of belief) and specific IoT environments (distributed, heterogeneous, nonlinear and object tracking environments).

Tribeni Prasad Banerjee and S. Das (Banerjee and Das, 2012) propose and investigate a hybrid method for fault signal classification based on sensor data fusion by using the Support Vector Machine (SVM) and Short Term Fourier Transform (STFT) techniques.

Othman Sidek and S. A. A. Quadri (Sidek and Quadri, 2012) review data fusion models, as well as aspects of systems engineering related to multisensory fusion. They propose a novel generic framework to link data fusion system engineering with algorithm engineering paradigm.

Mohamed Mostafa Fouad et al. (Fouad et al., 2015) give an overview about the state-of-the art of the data mining and data fusion techniques designed for the WSNs. It discusses how these techniques can prepare the sensor generated data in the network before any further processing as big data.

Xiaotie Qin and Y. Gu (Qin and Gu, 2011) discuss data fusion based on TGA.

RamezElmasri et al. (Lee et al., 2010) give an overview of fusion-related concepts along with their pros and cons. Eduardo F. Nakamura et al. (Nakamura, Loureiro and Frery, 2007) discuss techniques such as filters, Bayesian and Dempster–Shafer inference, aggregation functions, interval combination functions and classification methods for wireless sensor data fusion. Li Zhang and H. Gao (Zhang and Gao, 2017) propose a multi-sensor data fusion method based on deep learning for ball screw. Parallel superposition on frequency spectra of signals is directly done in the proposed deep learning-based multi-sensor data fusion method, and deep belief networks (DBN) are established by using fused data to adaptively mine available fault characteristics and automatically identify the degradation condition of ball screw.

Jemin George and L. Kaplan (George and Kaplan, 2011) propose an approach to solve the nonlinear least squares problem that arises in decentralised fusion. They further state that the additive nature of the measurement noise is lost when the signal is processed at the sensor node. The proposed approach employs the unscented transformation before the estimation. Shilpa Gite and H. Agrawal (Gite and Agrawal, 2016) discuss IoT middleware using context-aware mechanism. A review on context awareness for Multi Sensor Data Fusion in IoT is presented.

A summary of each relevant survey paper related to healthcare data fusion is given in Table 9.2.

TABLE 9.2

Major Data Fusion Surveys in the Healthcare Domain

Survey	Objective Summary
Hamid Medjahed et al. 2011)	Focus on a data fusion approach based on fuzzy logic with a set of rules directed by medical recommendations used to fuse the various subsystem outputs.
Tejal Shah et al. 2016)	QoS and resource management issues in data fusion are discussed. Workflow of remote health-monitoring in IoT is presented. Big data tools not capable of heterogeneous data collection, real-time patient monitoring and automated decision support (semantic reasoning) based on well-defined QoS constraints.
Rachel C. King et al. 2017)	In this paper data fusion techniques and algorithms that can be used to interpret wearable sensor data in the context of health monitoring applications are discussed.
Vangelis Metsis et al. 2012)	Machine learning framework for brain tumor classification based on heterogeneous data fusion of metabolic and molecular datasets, including state-of-the-art high-resolution magic angle spinning proton magnetic resonance spectroscopy and gene transcriptome profiling, obtained from intact brain tumor biopsies.
Jieping Ye et al. 2008)	A novel framework to integrate heterogeneous data for Alzheimer's disease prediction based on a kernel method is proposed. Further the kernel framework is extended for selecting features (biomarkers) from heterogeneous data sources.
Hyun Lee et al. (2009)	Important issues in healthcare monitoring with a focus on software problems such as reliability, network robustness and context awareness are discussed. New architecture for handling data cleaning, data fusion and context and knowledge generation using multi-tiered communication is proposed.

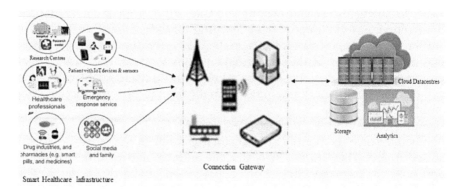

Smart Healthcare Infrastructure

FIGURE 9.3
Data fusion in IoT: Smart healthcare perspective.

9.3 Healthcare Data Fusion: Opportunities and Challenges

Figure 9.3 gives a brief overview of the smart healthcare infrastructure connection gateway, which includes low-level computational devices such as mobile phones belonging to in-network sensor data processing and high-end computational devices such as servers which belong to cloud level processing.

9.3.1 Healthcare Data Fusion: Opportunities

Data fusion needs to be used in healthcare data because of the following benefits:

1. Data fusion is essentially an information integration problem. It integrates data from multiple sensors to provide better analysis and decision-making in a situation that can be done using any single sensor.

2. Different types of sensors have different strengths and weaknesses. Therefore, integrating data from multiple sensors of different types provides a better result because the strengths of one type can compensate for the weaknesses of another type.

3. The availability of multiple heterogeneous datasets presents new opportunities to big data analysts, because the knowledge that can be acquired from combined data cannot be acquired from any individual source alone.

4. These quantitative data collectively can be used to do clinical text mining, predictive modelling (Wan et al., 2013), survival analysis, patient similarity analysis (Duan, Street and Xu, 2011), and clustering, to improve care treatment and reduce waste.

5. In healthcare area, data fusion can help association analysis, clustering and outlier analysis (Bellazzi and Zupan, 2008).

6. Fusing the healthcare data from heterogeneous sources can also be used to identify and understand high-cost patients (Silver et al., 2001) and applied to the mass of data generated by millions of prescriptions, operations and courses of treatment to identify unusual patterns and uncover fraud.

7. Data fusion also helps to build a context-awareness model that helps to understand situational context (Wang et al., 2015).

9.3.2 Healthcare Data Fusion: Challenges

Healthcare data fusion has multiple challenges, which are explored in various literature. Despite the technological advancements in big data processing technologies like Hadoop, Spark etc. and scalable infrastructure (e.g. clouds) various issues like data residing at different sources, inconsistent variable definitions, complexity of data, heterogeneous data collection, reliability, patient identification, data management, context and knowledge generation and real-time patient monitoring are identified. Some of them are listed below:

1 Data residing issue: From different source systems, like EMRs or HR software, to different departments, like radiology or pharmacy, the data comes from all over the organisation. Aggregating this data into a single, central system makes this data accessible and actionable. Healthcare data also occurs in different formats (e.g., text, numeric, paper, digital, pictures, videos, multimedia etc.). Sometimes the same data exists in different systems and in different formats.

2 Inconsistent/variable definitions: Healthcare data can have inconsistent or variable definitions. For example, one group of clinicians may define a unit of asthmatic patients differently than another group of clinicians. Two different clinicians have different criteria to identify someone as a diabetic. Knowledge of metrics when trying to capture health data is very important.

3 Data complexity: Claims data has been around for years and thus it has been standardised and scrubbed. But this type of data is incomplete. Clinical data from sources like EMRs give a more complete picture of the patient's story. These imperfections in the data must be dealt with effectively with the use of data fusion algorithms.

4 Reliability issues: Reliability in a healthcare application is critical
   because the system has to report correct data in a timely manner
   to the doctor or the healthcare provider. These reliability issues can
   be classified into three main categories (Lee et al., 2008): reliable
   data measurement, reliable data communication and reliable data
   analysis.

   a. Reliable data measurement: Internally implanted sensors, such
       as a retina or cortical prosthesis, can heat the surrounding tis-
       sues by generating radiation from wireless communication and
       power dissipation of sensors (Tang et al., 2005). This thermal
       effect by tissue heating can cause incorrect data measurement
       from the sensors and may harm the tissues. In order to mini-
       mise this effect, Tang et al. (Tang, Tummala and Gupta, 2005)
       proposed a rotating leadership algorithm within a cluster that
       considers the leadership history and the sensor locations. For the
       reliable measurement of external sensors, an extra redundant
       sensor (Lee et al., 2008) for the same function in case of a mal-
       function can be added.

   b. Reliable data communication: Most healthcare monitoring sys-
       tems adopt wireless communication, since it is useful to gather
       data anytime and anywhere. But, due to the intrinsic properties
       of wireless medium, such as signal attenuation and distortion,
       it is not reliable to transmit data in wireless networks. The data
       or packet delivery rate is even decreased when the network uses
       multi-hop communication, such as ad hoc wireless networks or
       wireless sensor networks. One of the possible solutions for reli-
       able data communication is to have multiple wireless networks,
       such as wireless sensor networks, ad hoc wireless networks,
       cellular networks, satellite networks and wireless LANs, as pro-
       posed by Varshney (2007). This architecture can utilise character-
       istics of each network, such as in terms of bandwidth, coverage,
       required power level, and priorities for access and transmission,
       whenever the characteristic is necessary. But, for architecture
       with non-homogeneous networks, specific hardware and proto-
       cols are needed to switch from one network to the other.

   c. Reliable data analysis: Once the data are measured and transmit-
       ted to a server correctly, the system has to analyze the data so that
       it can generate appropriate context. For reliable data analysis, the
       reliability in context generation is important so that the result of
       it can assist medical doctors to make the correct decision.

5 Identification of patients: Identification of targeted person and variable sensors provides more reliable and accurate healthcare monitoring, which is based on wireless communication. Proper patient
   identification is mandatory to avoid ambiguity in the source of the

sensed data and to classify the collected data depending on identification of a person. Identification is also needed in the localisation
and tracking of the patient. In Lee et al. (2008) the authors have proposed a multi-session-based passive High Frequency RFID system
to achieve better reliability and accuracy of identification.

6 Data management: Data management is a crucial aspect in the
Internet of Things. When considering a world of objects interconnected and constantly exchanging all types of information, the
volume of the generated data and the processes involved in the handling of those data become critical.

7 Data cleaning and summarising of data: In healthcare monitoring
system, accuracy of sensed data from patients is very important. If
uncertain data from the sensors is obtained, it causes a critical situation for the patient's well-being. Unfortunately, sensors have several
factors that affect data accuracy, such as environment effects and
hardware problems. The place for data cleaning is divided into two
locations: the sensor and the base station. Many researchers proposed methods to reduce the uncertainty. In Elnahrawy and Nath
(2003), the Bayesian approach is used for combining prior knowledge of true sensor readings, the noise characteristics of sensors and
the observed noisy reading.

8 Storage of data: After a body sensor generates a measurement, it
transmits sensed data to the base station. After data cleaning and
summarising, it is stored in the database of the base station. The
database will store all kinds of data such as sensedbody/environ-
ment/location/time data. Thus, it will need a well-organised storage
system.

9.4 Evaluation Framework

The foundation of Smart Healthcare is built on intelligent, low-power, wirelessly connected medical devices (Dautov, Distefano and Buyya, 2019). This
section evaluates the healthcare data fusion framework based on the following metrics.

9.4.1 Middleware Architecture Type

Middleware is software that provides services to software applications that
can't be obtained from the operating system. Middleware manages the interaction between disparate applications across the heterogeneous computing platforms (https://web.archive.org/web/20120629211518/http:/www.mi

ddleware.org/whatis.html, no date). Middleware is usually built to address the common issues in application development such as heterogeneity, interoperability, security and dependability. Interoperability is very important in IoT healthcare systems. Healthcare systems produce a large amount of health data. This healthcare data is spread among public and private institutions, clinics, care centres, laboratories, etc. One of the big challenges in the medical area is to provide access to patients' clinical data, regardless of where they were generated (Ahmed et al., no date).

a. Middleware helps to achieve interoperability. In many cases most EHR systems do not network, hence there is a need to add another application that draws records from that EHR and communicates through a middleman to facilitate the exchange of health information. This provides interoperability. Additionally, it saves physicians the step of having to convert the information into an interface, as most middleware platforms pull the data into a format that is easily transmissible.

b. Middleware provides flexibility between departments. Specific departments in hospitals are interested in only a particular set of patient information. Instead of having to sort through the entirety of a patient's records stored on an EHR system, middleware platforms allow different departments to draw from the EHR and organise the information quickly into a format that serves their needs best.

c. Middleware allows for connection with mobile devices. Middleware platforms can be transmitted to smartphones and other mobile devices, allowing for physicians to receive alerts with patient information. Mobile access can enhance efficiency in hospitals, by eliminating the step where physicians must find a terminal, log in and review patient information before making a decision.

d. Middleware allows hospitals to adapt more gradually to new EHR functionality. Middleware and software-as-a-service can help provide some of that functionality without the cost burden on providers and demands on EHRs.

Carlos Andrew Costa Bezerra et al. (Andrew et al., 2016) discuss the HL7-based middleware for heterogeneous healthcare data exchange. Health Level-7 (HL7) is a set of international standards to transfer clinical and administrative data between software applications used by various healthcare providers.

9.4.2 Context Awareness

Context awareness is the ability of a system or system component to gather information about its environment at any given time and adapt behaviors

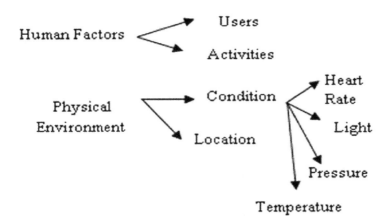

FIGURE 9.4
Context feature space.

accordingly. Contextual or context-aware computing uses software and hard-ware to automatically collect and analyze data to guide responses. Context information in healthcare applications can help to improve the quality of healthcare delivery, utilise limited healthcare and human resources more efficiently and to better match the healthcare services to the current medical conditions and needs of the patients under health monitoring (Bricon-Souf and Newman, 2007). Albrecht Schmidt et al. (Schmidt, Beigl and Gellersen, no date) state that context can be acquired either explicitly by requiring the users to specify it or implicitly by monitoring the user. So, based on the dis-ease, some features can be identified whose values will help in determining the context. Figure 9.4 shows context feature space.

Context information is needed in health monitoring to improve the quality of healthcare delivery, utilise the limited healthcare and human resources more efficiently and to move toward matching the healthcare services to the current medical conditions and needs of the patients under health monitoring. The context of a patient's current needs can be derived using the patient's medical history, current location and activity, values/rate of changes in the patient's current vital signs, among other pieces of informa-tion. Context-aware wireless health monitoring systems can be implemented and utilised to support diverse patients with a range of chronic conditions, using advances in devices, networks and protocols.

9.4.3 Semantic Interaction

IoT devices are highly heterogeneous in terms of underlying communication protocols, data formats and technologies, which makes it difficult to achieve interoperability in IoT systems. Also, lack of worldwide acceptable standards means that interoperability tools remain limited. In the healthcare domain physicians communicate with their patients with heterogeneous IoT devices to monitor their current health status. Information between physician and patient is semantically annotated and communicated in a meaningful way. In Pires, Garcia and Pombo (2016), a lightweight model for semantic annotation of data using heterogeneous devices in IoT is proposed to provide annotations for data. Resource Description Framework is used to provide Semantic Interoperability among physicians and patients using heterogeneous IoT devices. RDF annotated patients' data has made it semantically interoperable. SPARQL query is used to extract records from RDF graph. In Santos, Rodrigues, Silva and Denisov (2016), the authors used Intelligent Personal Assistant (IPA) as a software agent in IoT devices for physicians to give real-time information about the monitored patients.

A semantic model for description of smart objects using ontologies and description logics to enable Semantic Interoperability is proposed in Yachir et al. (2016). Jayaraman et al (Jayaraman et al., 2015) described an Open IoT platform used for digital agriculture use case. The Open IoT platform used ontologies to represent digital agriculture domain concepts to collect smart collection of information, annotation and validation processes to enable semantic interaction.

In Gyrard and Serrano (2016) the authors proposed SEG 3.0 methodology to unify, federate and provide Semantic Interoperability in IoT domain. The SEG 3.0 emerged from ontology engineering and its main benefit to integrate heterogeneous data collected from different smart things.

Pereira et al. (2016) proposed a design of ETSI M2M gateway integrated with libraries to enable interoperability between Machine-to-Machine (M2M) and IoT; common standards are the driving needs. In Practice et al., (2016), the authors proposed an architecture for data processing in IoT cloud environment that supports semantics interoperability. Google Cloud and Microsoft Azure were used as multi-cloud environments with Open IoT architecture.

In Shariatzadeh et al. (2016) digitising the factory in IoT is proposed. The author proposes integration of IoT and PLM platforms using semantic web technologies and Open Services for Lifecycle Collaboration (OSLC) standard on tool interoperability. Data communicates from heterogeneous IT environment to ensure data consistency. Despite a lot of research in semantic interoperability in the IoT healthcare domain security issues still remain unattended. It must be a critical aspect for the solution of interoperability.

9.4.4 Dynamic Configuration

Dynamic system configuration is the ability to modify and extend a system while it is running (Kramer, 1985). Dynamic configuration can be interpreted at two levels: software level and hardware level. In the healthcare scenario, failure of sensors cannot be afforded. Hence the healthcare network should be able to change, adapt and configure itself to the environment dynamically.

9.4.5 Fusion Complexity

In healthcare applications, there are situations when a single doctor does not have sufficient information and/or need to establish a diagnosis. In these cases, to diagnose a problem, a medical board of healthcare professionals with different specialisations needs to discuss and correlate their individual findings, after which the head of the board is able to take the final decision, based on the inputs provided by individual board members. From this perspective, this medical board resembles a hierarchical organisation, where lower-level members provide information to a higher-level decision-maker. These scenarios challenge as to how to process and manage large amounts of data, continuously generated by millions of sensing devices. Here data fusion comes into picture.

9.4.6 Actuation Management

Actuation is nothing other than acting according to the situation. Current healthcare systems have limited effectiveness in detecting emergency patient data and alerting the authorised medical caregiver. Currently most alert systems are configured as threshold alarm. Threshold alarm is generated when a vital sign crosses a threshold value. Control/actuation has two components: (a) decision-making and (b) mechanism.

For example, if fever is detected by a smart healthcare framework and this event is pushed to the controller, then the latter can choose to change the room temperature using actuation. This is an example of controlling smart home equipment based on eHealth sensor measurements. This is another unique aspect of the IoT architecture (Datta et al., 2015). Figure 9.5 gives an overview of actuation management in smart healthcare.

9.4.7 Data Processing Type

IoT-based healthcare system connects preconfigured devices for processing of data of patients to make smarter decision in a deadline-oriented fashion.

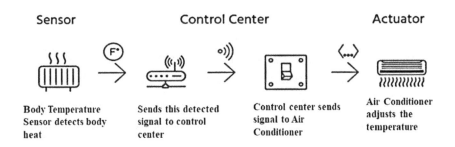

Sensor — Body Temperature Sensor detects body heat

Control Center — Sends this detected signal to control center / Control center sends signal to Air Conditioner

Actuator — Air Conditioner adjusts the temperature

FIGURE 9.5
Actuation management in smart healthcare.

Data processing in IoT can be done in two ways: in-network processing or cloud-level processing (Gill and Arya, 2019).

The healthcare data from heterogeneous IoT sensors has large volume and velocity. There is a need to utilise this big data to predict the current status of patients. After collection and aggregation of data from smart devices of IoT networks, data is stored and processed on cloud servers. To provide high scalable computing platform, the cloud performs the big data processing at infrastructure level (in-network). These platforms can be configured using pay-as-you-go-basis modes to fulfil the dynamic requirements of IoT applications, which decreases the cost and processes huge volumes of big data effectively using the cloud repository. The current cloud computing paradigm is unable to fulfil the dynamic requirements of IoT healthcare applications such as minimum response time and low latency. Therefore, a new paradigm, namely, fog computing has been introduced by Cisco (Gill and Arya, 2019), which fulfils the requirements of IoT healthcare applications and processes data within their specified deadline.

9.4.8 Cross Domain Portability

Healthcare organisations have a need to maintain their own files locally, which causes a geographic distribution of healthcare records. On the other hand, healthcare personnel treating a patient need access to previous diagnosis and treatment data, maintained by various institutions in many different locations (Graur, 2017). Hence there is a need for cross-domain portability in order that the health data can be exchanged. Currently, the lack of a reliable authentication and authorisation framework is considered a major obstacle for interchanging Electronic Healthcare Records (EHRs). One possible solution would be to develop a central medical database. But this approach has some disadvantages: (a) the privacy of patients and healthcare professionals and (b) the administrative burden of keeping such a central database is huge. Healthcare Information Systems (HISs) should ensure that only authorised people access the information. Many research efforts are leading the way in cross-domain portability for smart healthcare.

9.4.9 Implementation

Implementation of IoT application is critical in order to prove the IoT concept. Challenges that cannot been seen in theoretical processes can be seen during practical implementation. There are few things that must be considered before implementing an IoT healthcare application.

1. Safe collection of data must be ensured: In an IoT healthcare system collection of patient information is carried through special equipment (sensors etc.). This data collection is carried out outside the

network. Hence, when developing a healthcare application, it is very important to think of ways to protect the received information as the patient information needs to maintain privacy.

2. High-performance data streaming: Data collection systems consist of thousands of electronic devices. Hence, for the efficient streaming of such large volumes of information, it is necessary to think about independent mechanisms that are different from traditional packet transfers.

3. Internet of Things platform: The IoT platform is a set of software tools that collectively help to systematise, store and process data received from electronic devices.

4. Develop an Internet of Things solution in the cloud: For fast and efficient delivery of processed data to a user device, and also to organise centralised storage, cloud solutions are usually used. Such systems can ensure the efficient operation of the Internet of Things application with minimal operating costs and requirements for carrier networks.

5. Effective data management: In-memory analysis and data processing systems are mostly used. They ensure rapid delivery of the processed results to the end user, even in the event of the data collection device failure.

9.4.10 Performance Evaluation

Performance evaluation is an important issue in large-scale IoT devices. As the system gets larger, performance becomes a critical issue in IoT applications. In the healthcare domain it is the most important metric, which needs to be achieved effectively without any ambiguities. Surveys state that there is still a research gap in monitoring the performance of high-end IoT solutions (Wang et al., 2015).

9.4.11 Data Security and Privacy

With the increase in networked medical devices in healthcare, security attacks are growing. These security breaches need to be addressed and mitigated as they threaten to undermine technology development in the field and result in significant financial loses. A new report from the Atlantic Council and Intel Security, "The healthcare Internet of Things: Rewards and risks," says there is marked growth in adoption of these devices, with 48% of healthcare providers polled saying that they had integrated consumer technologies such as wearable health-monitoring devices or operational technologies like automated pharmacy-dispensing systems with their IT ecosystems. Once the device is compromised, the attacker can remotely manipulate devices, misuse medical records and control dosage levels of drugs. The sensitivity of data traveling in the IoT environment makes these types of attacks especially dangerous (Ragupathy and Thirugnanam, 2017).

9.5 Application of Data Fusion for Health Monitoring

This section discusses some applications of data fusion for health monitoring.

1. Brain tumor classification: Currently a lot of research is taking place in biomedical informatics that involves analysis of multiple hetero-geneous data sets. The data includes patient demographics, clinical and pathology data, treatment history, patient outcomes as well as gene expression, DNA sequences. Analysis of these data sets could lead to better disease diagnosis, prognosis, treatment and drug dis-covery. In (Metsis et al., 2012) a novel machine learning framework for brain tumor classification is presented. This framework integrates heterogeneous data sources like real biomedical/ biological MRS and genomic data. Later a combination of popular feature selection and classification methods is applied to evaluate the tumor-type dis-crimination capabilities of the two datasets separately and together. The feature selection process identifies a number of biomarkers from each dataset, which are used as features for the classification pro-cess. The experimental results show that this data fusion framework outperforms each individual dataset in the brain tumor multi-class classification problem. This framework can also be applied to any other biomedical and biological data fusion for sample classification and biomarker detection.

2. Gait analysis: Human walking contains important physiology, kine-matic and dynamic information. In real life, there are many appli-cations of human gait analysis such as monitoring the patient's recovery progress in clinical practice, the control strategy of bionic robots, etc. For example, stroke is a common neurodegenerative con-dition with a principal symptom of progressive limbs movement disorder. Therefore, objective and accurate inspection of gait param-eters is of great help to a neurologist for appropriate assessment and diagnosis of stroke patients (Qiu et al., 2018). Practically, it is criti-cal to develop an objective and quantitative approach to assess the patient's physical condition. Sensor data fusion algorithms can help calculate accurate gait parameters.

3. Physiological monitoring: Recently physiological signals have been widely used for the development of diagnosis support tools in

medicine. The use of multiple signals or physiological measures is carried out using data fusion techniques, which improves the accuracy of diagnostic care systems. In Uribe et al. (2018) the authors have proposed a physiological signal fusion architecture, based on the JDL model. This proposed architecture provides a more reliable diagnosis and treatment based on evidence to support the specialist in their decisions.

In Wireless Body Sensors (WBSN), biosensors collect periodically physiological measures and send them to the coordinator where the data fusion process takes place. However, processing the huge amount of data captured by the limited lifetime biosensors and taking the right decisions when there is an emergency are major challenges in WBSNs. In Habib and Makhoul (2016), the authors introduce a data fusion model using a decision matrix, an early warning score system and fuzzy set theory. This is achieved by combining raw data from various heterogeneous biosensor nodes. This helps to obtain information of greater quality and make accurate decisions about the situation of the patient based on the collected data, thus, allowing continuous monitoring of the health condition of the patient, identifying emergency situations and taking immediate and fast decisions by the coordinator of the WBSN.

4. Activity recognition: Activity recognition (AR) is a subtask in pervasive computing and context-aware systems. It detects the physical state of human in real-time (Habib and Makhoul, 2016). In recent years, there had been an exponential growth of AR technologies and much literature exists focusing on applying machine learning algorithms on obtrusive single modality sensor devices. Razaq et al. (Habib and Makhoul, 2016) present feature-level and decision-level fusion for multimodal sensors by pre-processing and predicting the activities within the context of training and test datasets.

In Palumbo et al. (2013)an Activity Recognition system that classifies a set of common daily activities is presented. In this work authors use both the data sampled by accelerometer sensors carried out by the user and the reciprocal Received Signal Strength (RSS) values coming from worn wireless sensor devices and from sensors deployed in the environment. Hence the proposed activity recognition system uses a mix of two approaches, i.e., wearable and not wearable. The accelerometer and the RSS stream that are obtained from a Wireless Sensor Network (WSN), using Recurrent Neural Networks, is implemented as efficient Echo State Networks (ESNs), within the Reservoir Computing paradigm. Finally, the results show that with an appropriate configuration of the ESN, the system reaches a good accuracy with a low deployment cost.

5. Fall detection and prediction: Fall detection and prediction is an intricate problem that involves complex interactions between physiological, behavioral and environmental factors.

   In Andò et al. (2016) the authors presented a multi-sensor data fusion approach, which fuses data from a gyroscope and an accelerometer. The authors have worked on smart algorithms for the activities of daily living (ADL) and fall classification. These algorithms utilise the data provided by inertial sensors embedded in a mobile phone, and installed on the user device. A threshold-based method applied to the features extracted from the average of the magnitude of the three acceleration and angular velocities components is used. The system automatically sends the notification to caregivers as soon as the fall event is detected. Wang et al. (Zhang, 2015) presented multi-sensor data fusion approach for fall prediction of older people. Simulation is carried on the walking assistant robot, which uses acceleration, gyroscope and tactile-slip sensors to acquire the patient's falling data and extracts its features. BP neural network algorithm is used for fall prediction.

6. Biomechanical modelling: The Kalman Filter–Parametric state estimation algorithm, which falls under the signal-level fusion method, can be used to measure biomechanical motions by combining accelerometer and gyroscope data to estimate the kinematic parameters. Due to the high power consumption of gyroscopes, other methods using multiple accelerometers are being developed, such as the double-sensor difference algorithm for the measurement of rotational angles of human segments.

7. Alzheimer's disease detection: In biomedical research, effective diagnosis of Alzheimer's disease (AD) is of primary importance. Recent studies have demonstrated that neuroimaging parameters are sensitive and consistent measures of AD. For this, integration of heterogeneous data (neuroimages, demographic and genetic measures) needs to be done. This can help to improve the prediction accuracy and enhance knowledge discovery from the data, such as the detection of biomarkers. In Ye et al. (2008) the authors propose to integrate heterogeneous data for AD prediction based on a kernel method. The kernel framework is further extended for selecting features (biomarkers) from heterogeneous data sources. The complementary voxel-based data and region of interest (ROI) data from MRI are the two data sources used in this study that integrate the complementary information by the proposed method. Experimental results show that the integration of multiple data sources leads to a considerable improvement in the prediction accuracy and also identifies biomarkers that play more significant roles than others in AD diagnosis.

9.6 Conclusion

This paper outlined the importance of data fusion in healthcare application. It describes some principles of data fusion and many of the foundation techniques that can be used to perform data fusion on healthcare big data. Further comparative analysis of major data fusion algorithms in healthcare domain is carried out. Various opportunities and challenges and recent applications of IoT healthcare using data fusion are explained. There is much scope to improve the efficiency of data fusion algorithms in the IoT big data environment for the healthcare domain.

References

Ahmed, E., et al., n.d., 'The role of big data analytics in Internet of Things', *Computer Networks*, 129, 459–471.

Akbar, A., et al., 2018, 'Real-time probabilistic data fusion for large-scale IoT applications', *IEEE Access*, vol. 6, pp. 10015–10027. doi: 10.1109/ACCESS.2018.2804623.

Alam, F., et al., 2017, 'Data fusion and IoT for smart ubiquitous environments: a survey', *IEEE Access*, vol. 5, pp. 9533–9554. doi: 10.1109/ACCESS.2017.2697839.

Andò, B., Baglio, S., Lombardo, C. O., Marletta, V., 2016, 'A multisensor data-fusion approach for ADL and fall classification', *IEEE Transactions on Instrumentation and Measurement*, vol. 65, no. 9, pp. 1960–1967. doi: 10.1109/TIM.2016.2552678.

Andrew, C., et al., 2016, 'Middleware for heterogeneous healthcare data exchange: a survey middleware for heterogeneous healthcare data exchange: a survey', In *ICSEA 2015: The Tenth International Conference on Software Engineering Advances*, pp. 409–414.

Ant, D. B., 2016, 'Sensor fusion algorithm based on extended Kalman filter for estimation of ground vehicle dynamics' In *IECON 2016-42nd Annual Conference of the IEEE Industrial Electronics Society*, pp. 1049-1054, IEEE. doi: 10.1109/IECON.2016.7793145.

Ashton, K., 2009, 'In the real world, things matter more than ideas', *RFID Journal*.

Bakr, M. A. and Lee, S., 2017, 'Distributed multisensor data fusion under unknown correlation and data inconsistency', *Sensors (Switzerland)*, vol. 17, no. 11. doi: 10.3390/s17112472.

Banerjee, T. P. and Das, S., 2012, 'Multi-sensor data fusion using support vector machine for motor fault detection', *Information Sciences*, vol. 217, pp. 96–107. doi: 10.1016/j.ins.2012.06.016.

Bareinboim, E. and Pearl, J., 2016, 'Causal inference and the data-fusion problem', *PNAS*, vol. 113, no. 27, pp. 7345–7352. doi: 10.1073/pnas.1510507113.

Bellazzi, R. and Zupan, B., 2008, 'Predictive data mining in clinical medicine: current issues and guidelines', *International Journal of Medical Informatics*, vol. 77, no. 2, pp. 81–97. doi: 10.1016/j.ijmedinf.2006.11.006.

Bricon-Souf, N. and Newman, C. R., 2007, 'Context awareness in health care: a review', *International Journal of Medical Informatics*, vol. 76, no. 1, pp. 2–12. doi: 10.1016/j.ijmedinf.2006.01.003.

Castanedo, F., 2013, 'A review of data fusion techniques', *ScientificWorld Journal*, vol. 2013, p. 704504. doi: 10.1155/2013/704504.

Datta, S. K., et al., 2015, 'Applying Internet of Things for personalized healthcare in smart homes', in *2015* 24th wireless and optical communication conference, WOCC 2015, pp. 164–169. doi: 10.1109/WOCC.2015.7346198.

Dautov, R., Distefano, S. and Buyya, R., 2019, 'Hierarchical data fusion for Smart Healthcare', *Journal of Big Data*. doi: 10.1186/s40537-019-0183-6.

Duan, L., Street, W. N. and Xu, E., 2011, 'Healthcare information systems: data mining methods in the creation of a clinical recommender system', *(B2) Enterprise Information Systems*, vol. 5, no. 2, pp. 169–181. doi: 10.1080/17517575.2010.541287.

Elnahrawy, E. and Nath, B., 2003, 'Cleaning and querying noisy sensors', in *Proceedings of the 2nd ACM international conference on wireless sensor networks and applications – WSNA '03*, p. 78. doi: 10.1145/941360.941362.

Fincher, D. W. and Mix, D. F., 1990, 'Multi-sensor data fusion using neural networks', in IEEE international conference on systems, man and cybernetics, 1990. Conference proceedings, pp. 835–838. doi: 10.1109/ICSMC.1990.142240.

Fouad, M. M., et al., 2015, 'Data mining and fusion techniques for WSNs as a source of the big data', *Procedia Computer Science*, vol. 65, pp. 778–786. doi: 10.1016/j.procs.2015.09.023.

George, J. and Kaplan, L. M., 2011, 'Multi-sensor data fusion: an unscented least squares approach', In *14th International Conference on Information Fusion*, no. 4, pp. 1–8.

Gibson, R. E., Hall, D. L. and Stover, J. A., 1994, 'An autonomous fuzzy logic architecture for multisensor data fusion', in *Proceedings of 1994 IEEE international conference on MFI 94 multisensor fusion and integration for intelligent systems*, vol. 43, no. 3, pp. 403–410. doi: 10.1109/41.499813.

Gill, S. S. and Arya, R. C., 2019, *Fog-based smart healthcare as a big data and cloud service for heart patients using IoT fog-based smart healthcare as a big data and cloud service for heart patients using IoT*, Springer International Publishing. doi: 10.1007/978-3-030-03146-6.

Gite, S. and Agrawal, H., 2016, 'On context awareness for multisensor data fusion in IoT', in: Satapathy S., Raju K., Mandal J., Bhateja V. (eds) *Proceedings of the Second International Conference on Computer and Communication Technologies. Advances in Intelligent Systems and Computing*, vol 381.

Graur, F., 2017, 'Dynamic network configuration in the Internet of Things', in *2017* 5th international symposium on digital forensic and security (ISDFS), pp. 1–4. doi: 10.1109/ISDFS.2017.7916503.

Gyrard, A. and Serrano, M., 2016, 'Connected smart cities: interoperability with SEG 3.0 for the Internet of Things', *30th International Conference on Advanced Information Networking and Applications Workshops (WAINA)*.

Habib, C. and Makhoul, A., 2016, 'Multisensor data fusion and decision support in wireless body sensor networks', *Conference: IEEE/IFIP Network Operations and Management Symposium, NOMS 2016*. doi: 10.1109/NOMS.2016.7502882.

ITU, 2005, 'The Internet of Things', Itu Internet Report *2005*, p. 212. doi: 10.2139/ssrn.2324902.

Jayaraman, P. P., et al., 2015, 'Do-it-yourself digital agriculture applications with semantically enhanced IoT building digital agriculture applications with semantically enhanced IoT platform', *2015 IEEE Tenth International Conference on Intelligent Sensors, Sensor Networks and Information Processing (ISSNIP)*. doi: 10.1109/ISSNIP.2015.7106951.

Jeff Kramer, J. M., 1985, 'Dynamic configuration for distributed systems', *IEEE Transactions on Software Engineering*, vol. 11, no. 4, pp. 424–436.

Junghans, M. and Jentschel, H.-J., 2007, 'Qualification of traffic data by Bayesian data fusion', in *Proceedings of the International Conference on Information Fusion*.

King, R. C., et al., 2017, 'Application of data fusion techniques and technologies for wearable health monitoring', *Medical Engineering and Physics*, vol. 42, pp. 1–12. doi: 10.1016/j.medengphy.2016.12.011.

Klein, L. A., 2004, *Sensor and data fusion*. Available at: http://www.mathgeek.net/_Media/spie.pdf.

Kumar, H. and Pimparkar, P., 2018, 'Data fusion for the Internet of Things', *International Journal of Scientific and Research Publications*, vol. 8, no. 3, pp. 278–282. doi: 10.29322/IJSRP.8.3.2018.p7541.

Lee, H., et al., 2008, 'Issues in data fusion for healthcare monitoring', in *Proceedings of the 1st ACM international conference on PErvasive Technologies Related to Assistive Environments - PETRA '08*, January 2008, p. 1. doi: 10.1145/1389586.1389590.

Lee, H., et al., 2010, 'Fusion techniques for reliable information: a survey', *International Journal of Digital Content Technology and its Applications*, vol. 4, no. 2, pp. 74–88. doi: 10.4156/jdcta.vol4.issue2.9.

Luprano, J., et al., 2007, 'Combination of body sensor networks and on-body signal processing algorithms: the practical case of MyHeart project', in International workshop on wearable and implantable body sensor networks (BSN'06), August, p. 4. doi: 10.1109/BSN.2006.15.

Medjahed, H., et al., 2011, 'A pervasive multi-sensor data fusion for smart home healthcare monitoring', in *2011 IEEE international conference on fuzzy systems (FUZZ-IEEE 2011)*, pp. 1466–1473. doi: 10.1109/FUZZY.2011.6007636.

Metsis, V., et al., 2012, 'Heterogeneous data fusion for brain tumor classification', *Oncology Reports*, vol. 28, no. 4, pp. 1413–1416. doi: 10.3892/or.2012.1931.

Nakamura, E. F., Loureiro, A. A. F. and Frery, A. C., 2007, 'Information fusion for wireless sensor networks', *ACM Computing Surveys*, vol. 39, no. 3, pp. 9. doi: 10.1145/1267070.1267073.

Paganelli, F., Spinicci, E. and Giuli, D., 2008, 'ERMHAN: a context-aware service platform to support continuous care networks for home-based assistance', *International Journal of Telemedicine and Applications*, vol. 2008. doi: 10.1155/2008/867639.

Palumbo, F., et al., 2013, 'Multisensor data fusion for activity recognition based on reservoir computing multisensor data fusion for activity', *Communications in Computer and Information Science*, vol. 386, no. 2. doi: 10.1007/978-3-642-41043-7.

Pereira, C., Pinto, A., Aguiar, A., Rocha, P., Santiago, F., Sousa, J., 2016, 'IoT interoperability for actuating applications through standardised M2M communications', *Proceedings of the 17th international symposium on a world of wireless, mobile and multimedia networks (WoWMoM '16)*.

Pires, I. M., Garcia, N. and Pombo, N., 2016, 'From data acquisition to data fusion: a comprehensive review and a roadmap for the identification of activities of daily living using', *Sensors*, vol. 16, no. 2, pp. 184. doi: 10.3390/s16020184.

Practice, S., et al., 2016, 'Analytics-as-a-service in a multi-cloud environment through', *Software: Practice and Experience*. doi: 10.1002/spe.

Qin, X. and Gu, Y., 2011, 'Data fusion in the Internet of things', *Procedia Engineering*, vol. 15, pp. 3023–3026. doi: 10.1016/j.proeng.2011.08.567.

Qiu, S., et al., 2018, 'MEMS inertial sensors based gait analysis for rehabilitation assessment via multi-sensor fusion', *Micromachines (Basel)*, vol. 9, no. 9, pp. 1–17. doi: 10.3390/mi9090442.

Ragupathy, S. and Thirugnanam, M., 2017, 'IoT in healthcare : breaching security issues security breaches and threat prevention in the Internet of Things'.

Rodríguez-Valenzuela, S., et al., 2014, 'Distributed service-based approach for sensor data fusion in iot environments', *Sensors (Switzerland)*, vol. 14, no. 10, pp. 19200–19228. doi: 10.3390/s141019200.

Santos, J., Rodrigues, J. J. P. C., Silva, B. M. C., Casal, J., Saleem, K. and Denisov, V., 2016, 'An IoT-based mobile gateway for intelligent personal assistants on mobile health environments', *Journal of Network and Computer Applications*, vol. 71, pp. 194–204.

Sasiadek, J. Z. Z. and Hartana, P., 2000, 'Sensor data fusion using Kalman filter', in *Proceedings of the third international conference on information fusion, FUSION 2000*, pp. 941–952. doi: 10.1109/IFIC.2000.859866.

Schmidt, A., Beigl, M. and Gellersen, H., n.d., 'There is more to context than location', *Computers and Graphics*, pp. 1–10.

Shah, T., et al., 2016, 'Remote health care cyber-physical system: quality of service (QoS) challenges and opportunities', *IET Cyber-Physical Systems: Theory and Applications*, vol. 1, no. 1, pp. 40–48. doi: 10.1049/iet-cps.2016.0023.

Shariatzadeh, N., et al., 2016, 'Integration of digital factory with smart factory based on Internet of Things', *Procedia CIRP*, vol. 50, pp. 512–517. doi: 10.1016/j.procir.2016.05.050.

Sidek, O. and Quadri, S. A. A., 2012, 'A review of data fusion models and systems', *International Journal of Image and Data Fusion*, vol. 3, 1, pp. 3–21. doi: 10.1080/19479832.2011.645888.

Silver, M., et al., 2001, 'Case study: how to apply data mining techniques in a healthcare data warehouse', *Journal of Healthcare Information Management*, vol. 15, no. 2, pp. 155–164.

Sundmaeker, H., Guillemin, P. and Friess, P., 2010, *Vision and challenges for realising the Internet of Things*. doi: 10.2759/26127.

Tang, Q., et al., 2005, 'Communication scheduling to minimize thermal effects of implanted biosensor networks in homogeneous tissue', *IEEE Transactions on Biomedical Engineering*, vol. 52, no. 7, pp. 1285–1294. doi: 10.1109/TBME.2005.847527.

Tang, Q., Tummala, N. and Gupta, S. K. S., 2005, 'TARA: thermal-aware routing algorithm for implanted sensor networks', in *Proceedings of 1st IEEE international conference distributed computing in sensor systems*, pp. 206–217. doi: 10.1007/11502593_17.

Thomas, C., 2010, *Sensor Fusion and Its Applications*. Available at: http://www.intechopen.com/books/sensor-fusion-and-its-applications.

Uribe, Y. F., et al., 2018, 'Physiological signals fusion oriented to diagnosis – a review physiological signals fusion oriented to diagnosis – a review', *Communications in Computer and Information Science*, vol. 885, pp. 1–15. doi: 10.1007/978-3-319-98998-3.

Varshney, U., 2007, 'Pervasive healthcare and wireless health monitoring', *Mobile Networks and Applications*, vol. 12, no. 2–3, pp. 113–127. doi: 10.1007/s11036-007-0017-1.

Vermesan, O. and Friess, P., 2013, *Internet of Things: converging technologies for smart environments and integrated ecosystems, challenges.* doi: 10.2139/ssrn.2324902.

Wan, J., et al., 2013, 'Cloud-enabled wireless body area networks for pervasive healthcare', *IEEE Network Magazine*, vol. 27, no. 5, pp. 56–61. doi: 10.1109/mnet.2013.6616116.

Wang, M., et al., 2015, 'City data fusion: sensor data fusion in the Internet of Things', *International Journal of Distributed Systems and Technologies (IJDST).* doi: 10.4018/IJDST.2016010102.

What Is Middleware? n.d., Available at: https://web.archive.org/web/20120629211518/http:/www.middleware.org/whatis.html.

Wikipedia.org., n.d., 'Data fusion', Available at: https://en.wikipedia.org/wiki/Data_fusion.

Wu, H., et al., 2002, 'Sensor fusion using Dempster-Shafer theory', *IEEE Instrumentation and Measurement Technology Conference*, vol. 12, pp. 21–23. doi: 10.1109/IMTC.2002.1006807.

Yachir, A., et al., 2016, 'A comprehensive semantic model for smart object description and request resolution in the internet of things', *Procedia - Procedia Computer Science*, vol. 83, pp. 147–154. doi: 10.1016/j.procs.2016.04.110.

Yaqoob, I., et al., 2016, 'Information fusion in social big data: foundations, state-of-the-art, applications, challenges, and future research directions', *International Journal of Information Management.* doi: 10.1016/j.ijinfomgt.2016.04.014.

Ye, J., et al., 2008, 'Heterogeneous data fusion for alzheimer's disease study', in *Proceeding of the 14th ACM SIGKDD international conference on knowledge discovery and data mining*, August, pp. 1025–1033. doi: 10.1145/1401890.1402012.

Yerva, S. R., Hoyoung, J. and Aberer, K., 2012, 'Cloud based social and sensor data fusion', in *2012 15th international conference on information fusion (FUSION)*, pp. 2494–2501.

Zhang, L. and Gao, H., 2017, 'A deep learning-based multi-sensor data fusion method for degradation monitoring of ball screws', in *Proceedings of 2016 prognostics and system health management conference, PHM-Chengdu 2016*, pp. 1–6. doi: 10.1109/PHM.2016.7819792.

Zhang, X., 2015, 'An approach for fall detection of older population based on multi-sensor data fusion', in *2015 12th international conference on ubiquitous robots and ambient intelligence (URAI)*, pp. 320–323. doi: 10.1109/URAI.2015.7358963.

Zheng, Y., 2015, 'Methodologies for cross-domain data fusion: an overview', *IEEE Transactions on Big Data*, vol. 1, no. 1, pp. 16–34. doi: 10.1109/TBDATA.2015.2465959.

10

Discriminative and Generative Model Learning for Video Object Tracking

Vijay K. Sharma, K. K. Mahapatra and Bibhudendra Acharya

CONTENTS

10.1 Introduction: Artificial Intelligence and Computer Vision

Artificial intelligence (AI) is the dominating area in industry as well as in academia. Information extraction from images, videos and texts, classification of data, clustering, expert systems, language translation, and face recognition are some of the important fields of AI.

The foundation of AI was laid down in the Dartmouth workshop organised by John McCarthy at Dartmouth College in Hanover, New Hampshire in 1956. It should be noted that the microprocessor is the heart of all computing systems and the first commercial microprocessor, Intel 4004, was available in 1971. AI is a computational system that involves massive computations and requires huge amount of data. It is because of this that applications related to AI were not widely available in the past. However, with the advancement of VLSI technology and networking systems we

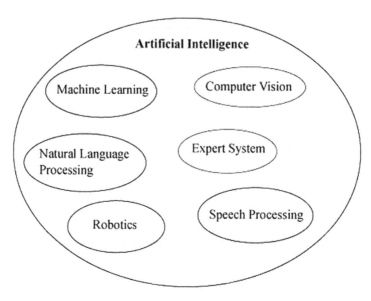

FIGURE 10.1
Core areas of artificial intelligence.

now have graphics processing units (GPU) that have thousands of cores (equivalent to a processing element) on a single chip and can process data on all cores in parallel. Advancement in communication and networking technologies has enabled the collection of data for AI system from different sources. The GPU technology has inspired the rapid development of applications based on AI.

The goal of artificial intelligence is to make a machine that is as intelligent as a human being. With a massive computational unit, an intelligent machine can process data and make a summary from huge collection of files. These are beyond the limit of an intelligent human. The self-driving car is the best example of progress taking place in AI. The major fields of AI are shown in Figure 10.1.

10.2 Computer Vision

Computer vision encompasses very important and interesting areas of AI. Images are the main component of the computer vision system. Information gathered through images are used by a system to take a specific action. A system captures images through cameras, analyses it through computer vision algorithms and performs certain task. For example, a driverless car senses the vehicles as well as the road in front of it through images captured by cameras attached to it, in order to steer in certain directions or apply the brakes (if there is an obstacle). Other important applications of computer vision are:

- Defect detections in various products
- 3-D reconstruction
- Object detection and classification
- Video object tracking
- Segmentation
- Image reconstruction
- Face detection and recognition
- Surveillance
- Optical character recognition
- Augmented reality

Typical steps in computer vision include image acquisition through sensors (cameras), pre-processing of images to enhance quality, feature extraction (e.g., edges, histograms), processing specific to the task (e.g., segmentation, classifier construction, etc.) and decision-making.

10.3 Introduction to Video Object Tracking

In video object tracking, the location and bounding box of an object is estimated as the object moves in the video sequence from one frame to the next. Video object tracking, also known as visual object tracking, is used in computer vision for the following applications:

- Hand tracking for gesture recognition
- Human computer interaction
- Autonomous vehicle driving
- Automated surveillance
- Traffic monitoring
- Action recognition
- Analysis of player strategies in sports
- Image compression

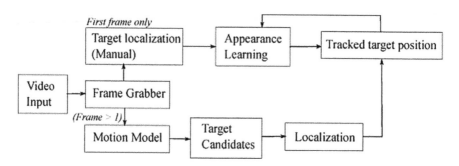

FIGURE 10.2
Steps in video object tracking.

The object tracking task becomes challenging because of the presence of the following factors. The visual appearance of the target object changes due to pose change, presence of occlusion, motion blur and deformation when the object moves across the video frames. Brightness variations, the presence of similar object in the surrounding of the target (called clutter) and target going out of camera view, fully or partially, are some other challenges in video object tracking. Therefore, an object tracking method must be robust so as to overcome these challenges.

Video object tracking has the following three major steps:

1. Appearance model
2. Motion model
3. Localisation

The purpose of learning an appearance model is to have a unique representation of the target that can be distinguished from many other objects. Appearance of the target can be constructed using either a feature descriptor or a mathematical model. The purpose of the motion model is to generate probable target candidates. In an image, there can be many patches equal in size to that of the target patch. The motion model tries to reduce the number of patches in order to reduce the computation. The localisation method chooses the correct target from the target candidates. After the target is localised, the location or the patch at the location is used to further update the target appearance model.

The steps involved in video object tracking are shown in a block diagram in Figure 10.2. From the first frame of the input video, the target is localised manually. This localised target is further used to construct an appearance model of the target. A frame grabber converts real-time video into image frames. The localised target from the second frame onwards is used to update the target appearance. Because there are several challenges related to target appearance, it is the most important component in video object tracking.

10.4 Appearance Model of the Target

Feature descriptors, such as scale invariant feature transform (SIFT), speeded-up robust features (SURF), histogram of oriented gradients (HOG), etc., can contain the appearance of the visual target (Zhou et al. 2009; Sun and Liu 2011; Qi et al. 2012; Li et al. 2012; He et al. 2009). The tracking is performed by matching the feature descriptor of the target with the feature descriptor of the target candidates obtained from the motion model. However, effective target appearance can be learned using statistical learning techniques to create a mathematical model as shown in Figure 10.3. Statistically learned mathematical appearance models can be classified as generative, discriminative and hybrid generative-discriminative.

10.4.1 Construction of Generative Appearance Model

In construction and learning of a generative appearance model, tracked visual objects in the successive video frames are used. One of the important and most popular generative appearance models is based on eigenbasis vectors (Ross et al. 2008). With some initial tracked samples, eigenbasis vectors based subspace is constructed using singular value decomposition (SVD). The learning of subspace based appearance (i.e., update of appearance) is performed when there are five new samples (batch size) available. This learning technique is called incremental subspace learning and is based on sequential Karhunen–Loeve algorithm (SKL) (Levey and Lindenbaum 2000). Trackers that use this incremental appearance learning technique are present (Cruz-Mota et al. 2013; Bai & Li 2012; Wang 2013). The basic tracking criterion is that we find out the reconstruction error of target candidates in the learned subspace (represented by eigenbasis vectors). The target candidate that has minimum reconstruction errors is the most probable target in the current frame.

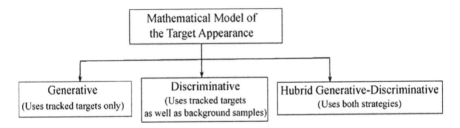

FIGURE 10.3
Appearance model in video object tracking.

10.4.2 Generative Appearance Model

The generative appearance model has high accuracy but it fails most of the time when the background is complex. The background component is recognised as a target when the target itself undergoes variations (e.g., deformation, rotation, etc.). The discriminative appearance model, therefore, considers background information also in order to learn a model that is able to discriminate the target from the background. In other words, the discriminative appearance model is based on a classifier that learns to classify a sample as target or background. A tracker based on support vector machine (SVM) is the best example of a discriminative tracker. The SVM is a binary classifier that constructs an optimal separating hyperplane between two sets of examples (positive and negative).

SVM is a supervised learning technique that requires labeled training data for learning. Suppose we have n number of samples in training data set $\{\mathbf{x}_i\}_{i=1}^n$, where each sample $\mathbf{x}_i$ is in $\mathbb{R}^d$ and the samples have the labels $y_i \in \{+1,-1\}$. The SVM learning tries to solves the following optimisation problem in order to get the maximum margin separating hyperplane ($\mathbf{w}^T\mathbf{x}+b=0, \mathbf{w}\in\mathbb{R}^d$ and $b\in\mathbb{R}$) (Boser et al. 1992; Platt 1998; Chang and Lin 2011),

$$\min_{w,b,\xi}\frac{1}{2}\mathbf{w}^T\mathbf{w}+C\sum_{i=1}^n\xi_i \tag{10.1}$$

subject to $y_i(\mathbf{w}^T\mathbf{x_i}+b)\geq 1-\xi_i,$

$$\xi_i\geq 0, i=1,\dots,n$$

The vector $\mathbf{w}$ is known as a parameter vector, which is normal to the hyperplane. The variable is equal to zero if the hyperplane passes through the origin. ξ_i's are called slack variables and C is a regularisation constant. The optimisation problem in Equation (10.1) is known as soft margin SVM formulation in primal space. The dual formulation is expressed in terms of Lagrange multipliers α_i as,

$$\max_{\alpha}\sum_{i=1}^n\alpha_i-\frac{1}{2}\sum_{i=1}^n\sum_{j=1}^n\alpha_i\alpha_jy_iy_j\mathbf{x}_i^T\mathbf{x}_j \tag{10.2}$$

subject to $\max_{\alpha}\sum_{i=1}^n\alpha_iy_i=0, 0\leq\alpha_i\leq C$

The normal vector w is expressed as the weighted sum of some *nSV* (*nSV* $\ll n$) number of examples which are near to the boundary of separating hyperplane, given in Equation (10.3). These examples are called support vectors. The weights are the values of Lagrange multipliers $\alpha_i > 0$.

$$\mathbf{w} = \sum_{i=1}^{nSV} \alpha_i y_i \mathbf{x}_i \qquad (10.3)$$

If we want to test an example $\mathbf{x}_i$, classification function $f(\mathbf{x}_i)$ or SVM score is expressed by the Equation (10.4) as

$$f(\mathbf{x}_i) = \mathbf{w}^T \mathbf{x}_i + \mathbf{b} \qquad (10.4)$$

For video object tracking, suppose we have N number of target candidates $\{\mathbf{x}_i\}_{i=1}^{N}$. These target candidates are cropped based on the motion model. We classify a candidate as a target if its score computed using Equation (10.4) is highest amongst all candidates. The dot product $\mathbf{x}_i^T \mathbf{x}_j$ in Equation (10.2) can be replaced with a kernel for learning in high dimensional space. In that case, Equation (10.4) is also replaced with a kernel function (Chang and Lin 2011).

An SVM based tracker that works on finding the maximum classification score is presented in Avidan (2004) as a support vector tracker (SVT) for vehicle tracking. There are a number of state-of-the-art trackers based on SVM learning (Hare et al. 2011; Bai and Tang 2012; Bai and Li 2014; Zhang, Ma and Sclaroff 2014; Ning et al. 2016). The tracker in Hare et al. (2011) is based on structured output framework of SVM (known as STRUCK). The SVM based discriminative model learned in the first video frame needs to be updated in every video frame in order to capture the target appearance as well as the current background. When the SVM parameter is updated in each frame, the process is called sequential learning. There have been approaches for sequential learning in the past (Freitas et al. 1999; Bordes et al. 2009; Bottou 2012; Wang et al. 2012; Davy et al. 2006; Zhang, Ma and Sclaroff 2014; Ning et al. 2016).

Ning et al. (2016) express that if we represent a target using a high-dimensional feature, the performance is better. However, the computational complexity is very high when we apply structured output SVM based onto kernels. Therefore, if we can explicitly learn a parameter vector in linear space, the update of the parameter is less complex.

There are other discriminative trackers that do not use SVM formulation. Kernelised correlation (KCF) based tracker (Henriques et al. 2015) is based on fast correlation computation using fast Fourier transform (FFT). This method considers all possible target candidates near the previous tracked location for finding the target. The background around the target is used for correlation computation. The spatio-temporal context (STC) based tracker in Zhang et al. (2014) also is based on correlation computation using kernel. Deep features from pretrained convolution neural networks (CNNs) are used in correlation based trackers for effective tracking (Ma et al. 2015). Multiple instance learning based trackers (Babenko et al. 2011; Zhang and Song 2013; Zhang et al. 2013; Zhang et al. 2014) use ratio classifier and online feature selection algorithms to select a few features from a large feature pool. The selected features are discriminative with respect to the background.

10.5 Motion Model

The motion model is used to identify the possible locations of the object in the current video frame. If we simply crop the patches of size equal to the size of the target in the entire image frame, there will be huge number of patches. Evaluating all these patches for finding a patch that contains the target is s computationally demanding operation. This is where the model comes in to save the computation. There are three types of approach used in video object tracking for target candidate generation: sliding window, mean-shift and particle filter. In addition, KCF based trackers uses all possible patches within a range of the target in the previous frame. In video object tracking, an object moves very slightly from one frame to the next. Therefore, it is unwise to crop patches far away from the target in the previous frame. Sliding window either crops all patches or leaves some patches between two cropped patches to save the computation in evaluation. A mean-shift based motion model iteratively reaches near to the target based on the mean of the data in a window (Comaniciu et al. 2003).

Particle filter is the motion model technique used to significantly reduce the number of target candidates. Particle filter is also known as sequential Monte Carlo (SMC) approach (Arulampalam et al. 2002). The moving target can be represented by a state vector that contains information such as height, width, rotation and its translation parameter. The target state in the next frame can be obtained from the posterior distribution of the state parameter. However, the posterior distribution also needs an update in each frame when the new observations are available. To obtain an analytic expression to update the posterior distribution is complex in the case of non-linear and non-Gaussian data (Arulampalam et al. 2002, Doucet et al. 2001).

The particle filter approximates the posterior using set of random particles and weights (Arulampalam et al. 2002). The particle weights are proportional to the observation likelihood when the resampling is used in each sequential step (Fan et al. 2015). Consider the state at time t is represented by s_t, i-th particle of the state by s_t^i and x_t as an observation vector. The observation likelihood in the state s_t^i is represented as $p(\mathbf{x}_t \mid \mathbf{s}_t^i)$. The observation likelihood is a measure that indicates the closeness of the observation in the state. If there are several random states then, from each state, we obtain an observation, compute the likelihood using a likelihood function and estimate the state based on maximum likelihood value. A simplest likelihood function can be $p(\mathbf{x}_t^i \mid \mathbf{s}_t^i) = \exp(-\mid\mid \mu - \mathbf{x}_t^i \mid\mid^2)$, where μ is the mean of tracked patches and $\mathbf{x}_t^i$ is a patch cropped for state vector $\mathbf{s}_t^i$, and both are column stacked to form 1-d vector.

State transition density $p(\mathbf{s}_t \mid \mathbf{s}_{t-1})$, i.e., generation of the next state given the current state, is approximated by a Gaussian density function (Fan et al. 2015; Arulampalam et al. 2002,#; Wang et al. 2013; Ross et al. 2008). If the state is represented by six affine transformation parameters, $\mathbf{s}_t = (x_t, y_t, \theta_t, \alpha_t, \phi)$, corresponding to (x, y) translation, rotation angle, scaling, aspect ratio and skew, then n samples $\mathbf{s}_t^i$, $i = 1, 2, ..., n$ can be generated using normal distribution $p(\mathbf{s}_t \mid \mathbf{s}_{t-1}) = \mathrm{N}(\mathbf{s}_t ; \mathbf{s}_{t-1}, \psi)$, with each parameter of the state modeled independently around the previous state $\mathbf{s}_{t-1}$ using a Gaussian distribution with a diagonal covariance matrix $\psi = (\sigma_x^2, \sigma_y^2, \sigma_\theta^2, \sigma_s^2, \sigma_\alpha^2, \sigma_\phi^2)$, In video object tracking, for each generated state, an image patch is cropped (Ross et al. 2008).

10.6 Proposed Method of Online Parameter Learning

We propose a parameter learning technique based on kernel similarity. The inner product between two vectors can be computed in high-dimension space with the help of a kernel trick. There is no need for explicit mapping of each pattern to a high-dimension space, thus avoiding computational burden. In general, a kernel function takes two data patterns and computes the similarity (or dot product, or distance) between them in high-dimension space, also called the feature space, using implicit feature mapping. Functions that satisfy Mercer's condition can be used as a kernel function. The Gaussian kernel is one of them, and maps the data into infinite dimension space to compute the similarity. The proposed method is given in Algorithm 1. In this method, the parameter vector is updated using positive and negative examples near the boundary $(V_{t,k})$. For each vector of vector set $V_{t,k}$, the parameter is modified iteratively until kernel similarity of positive and negative examples closest to the boundary in the current frame simultaneously satisfy the low and high similarity value with the value in the previous frame. The updated parameter is the one that has a high score as well as a high margin.

K positive examples ($K = 20$ in this work) are cropped using standard deviation [1,1,0,0,0,0]. We choose $sf_i = 0.0005$, $Msf = 0.25$. A kernel based similarity score, *KSim*, is computed between parameter vector $\mathbf{w}_t$ in the t-th frame and $\mathbf{T}_t^{Tr}$ (tracked target in the t-th frame). Further update of $\mathbf{w}_{t+1}$ (the parameter learned for (t+1)–th frame) is done if *KSim* value is below a threshold (0.6 in this work). This may correspond to a situation when the target is partially occluded or has undergone partial deformation, in which case the similarity value will be low. To update the parameter $\mathbf{w}_{t+1}$, the following steps are repeated Ns times ($Ns = 25$ for this work).

1. Crop some negative examples at a distance away from the target.
2. Find a negative example $\mathbf{x}_i$ that gives a minimum norm when the parameter is modified using $\mathbf{W}_{t+1} \leftarrow \mathbf{W}_{t+1} + lr1 \times (\hat{\mathbf{T}}_t^{Tr} - \hat{\mathbf{x}}_i)$.
3. Using the same negative example, $\mathbf{x}_i$, the update, $\mathbf{W}_{t+1} \leftarrow \mathbf{W}_{t+1} + lr1 \times \left(\hat{\mathbf{T}}_t^{Tr} - \hat{\mathbf{x}}_i \right)$, continues till the norm of $\mathbf{w}_{t+1}$ decreases (search for minimum norm).

The value of $lr1$ is 0.01. The likelihood functions to crop the intermediate instances are very simple.

$$p_1(\mathbf{x}_t^i \mid \mathbf{s}_t^i) = \mathbf{w}_{t+1}^T \mathbf{x}_{t+1}^i \tag{10.5}$$

$$p_2(\mathbf{x}_t^i \mid \mathbf{s}_t^i) = \mathbf{Tg}_{t+1}^T \mathbf{x}_{t+1}^i \tag{10.6}$$

$$p_3(\mathbf{x}_t^i \mid \mathbf{s}_t^i) = \mathbf{w}_{t+1}^T \mathbf{x}_{t+1}^i + \mathbf{Tg}_{t+1}^T \mathbf{x}_{t+1}^i \tag{10.7}$$

where $\mathbf{Tg}_{t+1}$ is the learned generative model, given by,

$$\mathbf{Tg}_{t+1} = \mathbf{Tg}_t + KSim \times \mathbf{T}_{t+1}^{Tr} \tag{10.8}$$

$\mathbf{Tg}_1$ is constructed using a tracked sample in the first frame. Corresponding to each likelihood, one instance is cropped and reshaped to a 1-d vector after HOG feature extraction. The final estimated instance is obtained by comparing their score with a discriminative model $\mathbf{wM}^t$ learned in each frame. The score is computed by simple inner product ($(\mathbf{wM}_t)^T \mathbf{x}^i$), where x^i is HOG feature vector of i-th instance. The discriminative model $\mathbf{wM}^t$ is learned in the first video frame using LIBSVM software (Chang and Lin 2011). In the subsequent frame, $\mathbf{wM}^t$ is learned as,

$$\mathbf{wM}_{t+1} = \widehat{\mathbf{wM}}_t + lr \times (\widehat{\mathbf{Vp}}_{t+1} - \widehat{\mathbf{Vn}}_{t+1}) \tag{10.9}$$

where, lr is the learning rate (equal to 0.02) in this work, $\mathbf{Vp}_{t+1}$ is average of current tracked sample and some other samples cropped near to the tracked sample with standard deviation [2,2,0,0,0,0], and $\mathbf{Vn}_{t+1}$ is the average of two negative instances.

In order to reduce the target drift, 50 new target candidates are cropped with standard

----

Algorithm 1 Discriminative Parameter Learning Using Kernel Similarity

----

Input : SVM parameter in the t - th frame, $\mathbf{w}_t$; Minimum and Maximum kernel

similarities $MinKS_t$ and $MaxKS_t$; Initial value of scaling sf_i;

Maximum scaling factor Msf

1 : Crop K positive examples very close to current tracked location

2 : Extract the HOG features and reshape to 1 - d vector, $\{PosV_{t,k}\}_{k=1}^{K}$, where each

$PosV_{t,k} \in \mathbb{R}^d$

3 : Find the kernel similarity score of each $\widehat{PosV}_{t,k}$ with $\hat{\mathbf{w}}_t$ and arrange them in

descending order of scope

4 : Crop some negative samples at some distance away from target

5 : Extract the HOG features and reshape to 1 - d vector, $\{NegV_{t,k}\}_{k=1}^{K}$, where each

$NegV_{t,k} \in \mathbb{R}^d$

6 : Find the kernel similarity score of each $\widehat{NegV}_{t,k}$ with $\hat{\mathbf{w}}_t$ and arrange them in

ascending order of score

7 : $V_{t,1} = \widehat{PosV}_{t,1} - \widehat{NegV}_{t,1}$

8 : **for** $k = 2$ to K **do**

9 : $V_{t,k} = (\widehat{PosV}_{t,k} - \widehat{NegV}_{t,k}) + V_{t,k-1}$

10 : **end for**

11 : $\mathbf{p}1 = \widehat{PosV}_{t,1}$; $\mathbf{n}1 = \widehat{NegV}_{t,1}$

12 : **for** $k = 1$ to K **do**

13 : $sf = sf_i$

14 : $\mathbf{w}_t^1 = \mathbf{w}_t + sf \times \hat{V}_{t,k}$

15 : $K_p = \mathcal{K}(\mathbf{p}1, \hat{\mathbf{w}}_t^1)$

16: $K_n = \mathcal{K}(n1, \widehat{\mathbf{w}}_t^1)$

17: **while** $(K_p \leq MinKS_t)$ **and** $(K_n \geq MaxKS_t)$ **and** $(sf \leq Msf)$ **do**

18: $\mathbf{w}_t^1 = \mathbf{w}_t + sf \times \widehat{V}_{t,k}$

19: $K_p = \mathcal{K}(p1, \widehat{\mathbf{w}}_t^1)$

20: $K_n = \mathcal{K}(n1, \widehat{\mathbf{w}}_t^1)$

21: $sf = sf \times 1.5$

22: **end while**

23: $Scr = \sum\limits_{k=1}^{K}((\widehat{\mathbf{w}}_t^1)^{\mathrm{T}}\widehat{PosV}_{t,k} + (\widehat{\mathbf{w}}_t^1)^{\mathrm{T}}\widehat{NegV}_{t,k})$

24: $M = K_p - K_n$

25: $\mathbf{w}_{t,k} = \widehat{\mathbf{w}}_t^1$

26: $DS_k = Scr + M$

27: **end for**

28: $k^* = \arg\max\limits_{k} DS_k$

29: $\mathbf{w}_{t+1} = \mathbf{w}_{t,k^*}$

30: Find out the two vectors $\mathbf{v}_1$ and $\mathbf{v}_2$ corresponding to index k^* in positive and negative sets

31: Find out inner product of $\mathbf{v}_1$ and $\mathbf{v}_2$ with $\mathbf{w}_{t+1}$ to get the minimum positive $MinKS_{t+1}$ and maximum negative $MaxKS_{t+1}$ score, respectively for the next frame

Output : Learned parameter vector $\mathbf{w}_{t+1}$

deviation [15,15,0,0,0,0]. Discriminative scores of all 50 candidates and the tracked target in the current frame are computed with discriminative parameters learned in the first video frame, $\mathbf{w}_1$. If the score of the tracked target is below 0.5 and also if the highest score of the new target candidate is 1.5 times higher than the score of tracked target, then the new tracked target is the target candidate with the highest score.

10.7 Experimental Results

The kernel function used in proposed tracking method is Gaussian kernel given by,

$$\mathcal{K}\left(\mathbf{x}_1, \mathbf{x}_2\right) = \exp\left(\frac{-||\mathbf{x}_1 - \mathbf{x}_2||^2}{2\sigma^2}\right) \text{ for } \sigma > 0 \tag{10.10}$$

where the value of σ is set to 1. A total of 800 target candidates are cropped without scaling with standard deviation [12,12,0,0,0,0]. To get the initial parameter vector $\mathbf{w}_1$ in the first frame, the SVM training in dual space is performed using LIBSVM software (Chang and Lin 2011). The number of positive examples for SVM training is 200, while negative examples are approximately 350. In order to estimate the scale of the target, 25 new target candidates are cropped at the location of tracked target with standard deviation [0,0,0.015,0,0,0]. The estimated scale of the target is the scale of the candidate with maximum score computed with discriminative parameter $\mathbf{w}_1$ learned in the first video frame.

Simulations are done on an Intel core i7 Processor with 3.4 GHz frequency on Windows 10 platform with 8 GB of RAM and MATLAB 2016a. For quantitative evaluation, the two most popular choices are center location error (CE) and overlap rate (OR). Center location error measures the Euclidean distance between ground truth location of the target and the tracked location of the target. OR is defined as intersection and union of bounding boxes of the ground truth (BG) and tracked object (BT), (Wang et al. 2013), given by,

$$Score = \frac{area(B_T \cap B_G)}{area(B_T \cup B_G)} \tag{10.11}$$

We have tested our proposed method on CVPR 2013 dataset (Wu et al. 2013). The dataset contains video sequences where target undergoes low to severe rotation, scale change, occlusion, deformation and motion blur. The brightness of the target background also varies in most of the videos.

Comparative analysis is done with the following state-of-the-art video trackers: IVT (incremental visual tracking) (Ross et al. 2008); L-1 tracker based on sparse prototype (L-1 (SP)) (Wang et al. 2013); ODFS (Zhang et al. 2013); STC (Zhang et al. 2014); KCF (on HOG) (Henriques et al. 2015); MEEM (Zhang, Ma and Sclaroff et al. 2014); DLSSVM (with and without scale) (Ning et al. 2016); and CNN (Ma et al. 2015). We chose a unique set of parameters for all video sequences in all trackers for a fair evaluation of different trackers. The comparative results for center error and overlap rate are given in Table 10.1 and Table 10.2, respectively. NR in Table 10.1

TABLE 10.1

Average Center Error (CE) of Different Trackers

Video	IVT	L-1(SP)	ODFS	STC	KCF (HOG)	MEEM	DLSSVM	Scale-DLSSVM	CNN	Proposed
Boy	112.1	20.1	79.2	25.9	2.8	**2.5**	2.8	2.8	2.9	1.7
Car4	69.8	125.3	73.7	10.6	9.8	18.0	18.6	**3.8**	6.8	2.2
CarDark	1.5	7.7	32.2	2.8	6.0	1.2	1.4	**1.3**	5.8	1.2
CarScale	68.3	69.2	76.9	89.3	16.1	20.4	13.2	55.1	29.2	9.4
Coke11	123.8	124.4	50.9	74.3	18.6	10.9	8.7	6.3	10.4	22.2
Couple	69.8	10.3	9.4	NR	47.5	6.4	6.8	**5.9**	7.3	4.0
Crossing	27.6	2.7	8.6	34.0	2.2	**2.0**	**2.0**	2.7	2.7	1.5
David	222.5	5.6	63.9	12.1	8.0	9.3	7.9	2.8	7.7	**4.0**
David2	**1.3**	14.5	19.6	5.5	2.0	1.7	2.0	1.2	3.5	1.4
Deer	6.4	6.4	91.2	400.8	21.1	4.5	**4.3**	4.2	5.1	5.6
Dog1	3.1	6.2	6.4	21.1	**4.2**	6.0	5.1	5.9	4.6	10.0
Doll	23.3	30.5	12.0	NR	8.3	**4.2**	5.7	3.5	5.2	5.4
Dudek	9.0	**9.7**	27.4	25.5	12.0	14.9	12.6	10.2	10.7	10.0
Faceocc	17.2	**12.8**	23.2	250.4	13.9	16.4	15.9	17.7	19.4	12.6
Faceocc2	14.4	11.8	12.5	10.1	7.6	6.3	7.5	8.5	**7.0**	10.7
Fish	2.5	**2.6**	26.3	3.9	4.0	7.3	4.6	5.1	4.2	4.3
FleetFace	26.0	60.0	55.9	85.0	25.5	37.0	**23.4**	23.0	27.6	26.9
Football	182.1	153.0	15.9	10.0	5.8	4.7	5.2	3.4	4.6	**3.8**
Football1	14.1	7.8	22.0	72.5	5.4	5.1	5.1	7.5	2.6	**3.7**
Freeman1	79.1	12.9	62.1	NR	94.8	9.8	85.0	**8.3**	8.0	27.0
Freeman3	28.4	43.8	45.4	39.4	19.2	4.9	6.0	**3.7**	19.9	2.9
Girl	21.8	16.4	20.0	21.8	11.9	**3.8**	3.6	3.9	3.9	13.4

(Continued)

TABLE 10.1 (CONTINUED)

Average Center Error (CE) of Different Trackers

Video	IVT	L-1(SP)	ODFS	STC	KCF (HOG)	MEEM	DLSSVM	Scale-DLSSVM	CNN	Proposed
Jogging(1)	96.0	86.3	88.2	149.9	88.2	6.5	3.1	4.2	**3.8**	6.4
Jogging(2)	100.6	7.8	126.2	160.1	144.4	15.1	6.1	**4.3**	3.1	4.4
Jumping	4.5	4.4	99.1	67.2	26.1	4.4	4.6	4.5	**4.3**	3.8
Lemming	17.8	98.5	**13.3**	96.2	77.8	8.4	152.0	79.6	153.4	31.1
Liquor	549.2	302.9	182.8	113.4	90.5	11.9	8.1	**8.6**	22.9	13.1
Mhyang	1.3	2.4	30.6	4.5	3.9	4.9	2.7	2.3	2.5	**2.1**
Shaking	193.1	224.2	10.2	9.6	112.5	**7.9**	7.5	19.9	11.1	8.4
Singer1	8.2	**4.4**	18.7	5.7	10.7	49.7	15.4	5.2	9.2	3.4
Singer2	70.9	108.1	56.4	52.7	**10.2**	174.8	179.1	175.5	184.8	10.1
Subway	118.6	138.1	10.1	NR	2.9	4.2	3.1	2.4	**2.8**	3.2
Suv	62.1	53.4	89.4	51.4	3.4	36.5	16.0	89.8	5.1	**4.2**
Tiger1	72.4	75.1	105.7	63.5	**8.0**	13.5	10.7	14.8	11.7	**9.1**
Tiger2	95.2	50.5	53.9	57.4	47.4	19.6	16.6	**16.4**	19.2	12.6
Trellis	70.5	91.1	42.8	33.7	7.7	5.8	7.7	**3.6**	6.4	3.1
Walking	**2.0**	1.7	10.3	7.1	3.9	4.6	6.1	6.1	3.9	2.1
Walking2	78.3	74.2	65.1	13.8	28.9	22.8	18.9	2.5	9.4	**2.8**

TABLE 10.2

Average Overlap Rate (OR) of Trackers

Video	IVT	L-1(SP)	ODFS	STC	KCF (HOG)	MEEM	DLSSVM	Scale-DLSSVM	CNN	Proposed
Boy	0.182	0.612	0.241	0.543	0.777	**0.792**	0.779	0.782	0.774	*0.827*
Car4	0.313	0.292	0.170	0.351	0.483	0.457	0.467	**0.681**	0.492	*0.876*
CarDark	0.748	0.673	0.445	0.750	0.614	*0.861*	**0.858**	0.857	0.639	*0.760*
CarScale	**0.515**	0.494	0.358	0.423	0.419	0.413	0.423	0.476	0.417	*0.626*
Coke11	0.025	0.029	0.224	0.104	0.549	0.662	**0.729**	0.775	0.654	*0.561*
Couple	0.239	0.486	0.569	NR	0.200	0.599	**0.602**	**0.602**	0.586	*0.694*
Crossing	0.279	0.370	0.535	0.248	0.710	0.711	**0.731**	0.705	0.708	*0.775*
David	0.180	0.611	0.244	0.522	0.538	0.530	0.544	**0.764**	0.541	*0.770*
David2	0.800	0.606	0.219	0.588	0.827	**0.852**	0.835	*0.867*	0.753	*0.846*
Deer	0.744	0.728	0.285	0.040	0.623	0.748	0.752	*0.788*	0.746	**0.772**
Dog1	*0.749*	**0.732**	0.521	0.506	0.550	0.549	0.550	0.694	0.542	*0.706*
Doll	**0.666**	0.560	0.509	NR	0.534	0.568	0.568	0.660	0.566	*0.801*
Dudek	**0.782**	0.780	0.629	0.587	0.727	0.713	0.728	0.793	0.734	*0.777*
Faceocc	0.743	**0.796**	0.681	0.186	0.774	0.746	0.756	0.724	0.715	*0.797*
Faceocc2	0.552	0.534	0.685	0.689	*0.751*	0.775	**0.761**	0.734	0.759	*0.715*
Fish	0.783	0.807	0.457	0.510	0.839	0.764	0.824	0.807	**0.833**	*0.830*
FleetFace	0.548	0.527	0.554	0.419	0.589	0.566	**0.609**	0.619	0.580	*0.581*
Football	0.241	0.305	0.521	0.593	0.762	0.799	0.783	0.844	0.801	**0.828**
Football1	0.421	0.559	0.172	0.354	0.710	0.673	0.702	0.536	0.808	**0.788**
Freeman1	**0.388**	**0.441**	0.122	NR	0.214	0.393	0.217	0.407	0.422	*0.464*
Freeman3	0.388	0.373	0.002	0.251	0.324	0.342	0.346	0.335	0.313	*0.765*
Girl	0.160	0.207	0.255	0.334	0.545	**0.703**	0.728	0.691	0.702	*0.556*

(Continued)

TABLE 10.2

Average Overlap Rate (OR) of Trackers

Video	IVT	L-1(SP)	ODFS	STC	KCF (HOG)	MEEM	DLSSVM	Scale-DLSSVM	CNN	Proposed
Jogging(1)	0.171	0.179	0.172	0.168	0.185	0.667	**0.749**	0.710	0.724	0.775
Jogging(2)	0.123	0.653	0.092	0.143	0.124	0.585	0.724	**0.762**	0.776	0.731
Jumping	0.659	0.693	0.012	0.069	0.274	**0.706**	0.702	0.680	0.711	0.675
Lemming	0.378	0.389	0.626	0.231	0.384	0.680	0.227	0.340	0.230	**0.631**
Liquor	0.212	0.212	0.195	0.253	0.431	0.762	**0.801**	0.806	0.716	0.771
Mhyang	0.842	0.755	0.387	0.688	0.796	0.775	0.815	0.826	0.812	**0.830**
Shaking	0.008	0.008	0.661	0.623	0.039	**0.696**	0.721	0.484	0.659	0.721
Singer1	0.660	**0.790**	0.330	0.531	0.355	0.282	0.350	0.551	0.355	0.823
Singer2	0.230	0.123	0.474	0.408	0.732	0.043	0.043	0.042	0.039	**0.717**
Subway	0.147	0.153	0.524	NR	0.754	0.665	0.713	**0.779**	0.781	0.694
Suv	0.379	0.480	0.182	0.512	0.883	0.641	0.734	0.469	**0.816**	0.796
Tiger1	0.243	0.282	0.133	0.261	0.785	0.670	0.726	0.628	0.707	**0.756**
Tiger2	0.143	0.320	0.160	0.110	0.354	0.538	**0.599**	0.582	0.569	0.671
Trellis	0.223	0.257	0.361	0.470	0.631	0.619	0.621	0.813	0.612	**0.748**
Walking	**0.744**	0.768	0.497	0.598	0.530	0.537	0.543	0.546	0.543	0.684
Walking2	0.333	0.325	0.220	0.518	0.395	0.439	0.476	**0.716**	0.491	0.788

FIGURE 10.4
Tracked rectangles in Car4 video sequence.

FIGURE 10.5
Tracked rectangles in David video sequence.

and Table 10.2 is used to indicate no result could be obtained. Rectangles showing tracked object positions in the video frame are shown in Figure 10.4, Figure 10.5, Figure 10.6, Figure 10.7 and Figure 10.8. The proposed method is able to estimate the position as well as the size of the object in the video frame.

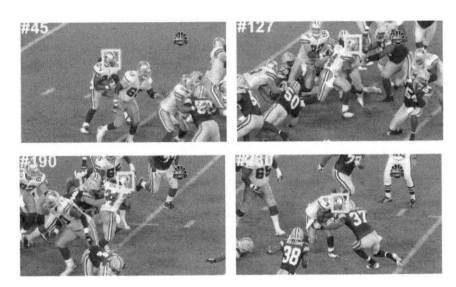

FIGURE 10.6
Tracked rectangles in Football video sequence.

FIGURE 10.7
Tracked rectangles in shaking video sequence.

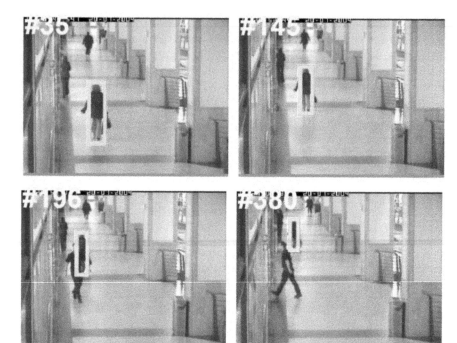

FIGURE 10.8
Tracked rectangles in Walking2 video sequence.

10.8 Conclusion

In this chapter, an online learning method is proposed for updating a parameter vector in the presence of sequential observations obtained in successive video frames. The method is based on finding a similarity score between parameter vectors and some examples that are near to the hyperplane. The method simultaneously maximises the score of some examples and margins using iterative learning steps. In a number of challenging video sequences, the proposed method performs better than several existing state-of-the-art trackers.

References

Arulampalam, M. S., Maskell, S., Gordon, N. and Clapp, T., 2002, 'A tutorial on particle filters for online nonlinear/non-gaussian bayesian tracking', *IEEE Transactions on Signal Processing*, vol. 50, no. 2, pp. 174–188.

Avidan, S., 2004, 'Support vector tracking', *IEEE Transactions on Pattern Analysis and Machine Intelligence*, vol. 26, no. 8, pp. 1064–1072.

Babenko, B., Yang, M.-H. and Belongie, S., 2011, 'Robust object tracking with online multiple instance learning', *IEEE Transactions on Pattern Analysis and Machine Intelligence*, vol. 33, no. 8, pp. 1619–1632.

Bai, T. and Li, Y., 2014, 'Robust visual tracking using flexible structured sparse representation', *IEEE Transactions on Industrial Informatics*, vol. 10, no. 1, pp. 538–547.

Bai, T. and Li, Y. F., 2012, 'Robust visual tracking with structured sparse representation appearance model', *Pattern Recognition*, vol. 45, no. 6, pp. 2390–2404.

Bai, Y. and Tang, M., 2012, 'Robust tracking via weakly supervised ranking svm', in IEEE conference *on* computer vision *and* pattern recognition *(CVPR)*, pp. 1854–1861.

Bordes, A., Bottou, L. and Gallinari, P., 2009, 'Sgd-qn: careful quasi-Newton stochastic gradient descent', *Journal of Machine Learning Research*, vol. 10, pp. 1737–1754.

Boser, B. E., Guyon, I. M. and Vapnik, V. N., 1992, 'A training algorithm for optimal margin classifiers', in *Proceedings of the Fifth Annual Workshop on Computational Learning Theory*, ACM, pp. 144–152.

Bottou, L., 2012, 'Stochastic gradient descent tricks', in Neural networks: tricks of the trade, Springer, pp. 421–436.

Chang, C.-C. and Lin, C.-J., 2011, 'LIBSVM: a library for support vector machines', *ACM Transactions on Intelligent Systems and Technology (TIST)*, vol. 2, no. 3, p. 27.

Comaniciu, D., Ramesh, V. and Meer, P., 2003, 'Kernel-based object tracking', *IEEE Transactions on Pattern Analysis and Machine Intelligence*, vol. 25, no. 5, pp. 564–577.

Cruz-Mota, J., Bierlaire, M. and Thiran, J., 2013, 'Sample and pixel weighting strategies for robust incremental visual tracking', *IEEE Transactions on Circuits and Systems for Video Technology*, vol. 23, no. 5, pp. 898–911.

Davy, M., Desobry, F., Gretton, A. and Doncarli, C., 2006, 'An online support vector machine for abnormal events detection', *Signal Processing*, vol. 86, no. 8, pp. 2009–2025.

Doucet, A., Freitas, N. D. and Gordon, N., 2001, 'An introduction to sequential monte carlo methods', in *Sequential Monte Carlo methods in practice*, Springer, pp. 3–14.

Fan, Z., Ji, H. and Zhang, Y., 2015, 'Iterative particle filter for visual tracking', *Signal Processing: Image Communication*, vol. 36, pp. 140–153.

Freitas, N. D., Milo, M., Clarkson, P., Niranjan, M. and Gee, A., 1999, 'Sequential support vector machines', in *Proceedings of the IEEE Signal Processing Society Workshop Neural Networks for Signal Processing IX*, pp. 31–40.

Hare, S., Saffari, A. and Torr, P. H., 2011, 'Struck: structured output tracking with kernels', in *IEEE International Conference on Computer Vision (ICCV)*, pp. 263–270.

He, W., Yamashita, T., Lu, H. and Lao, S., 2009, 'Surf tracking', in IEEE 12th *International Conference on Computer Vision*, pp. 1586–1592.

Henriques, J., Caseiro, R., Martins, P. and Batista, J., 2015, 'High-speed tracking with kernelized correlation filters', *IEEE Transactions on Pattern Analysis and Machine Intelligence*, vol. 37, no. 3, pp. 583–596.

Li, J., Wang, Y. and Wang, Y., 2012, 'Visual tracking and learning using speeded up robust features', *Pattern Recognition Letters*, vol. 33, no. 16, pp. 2094–2101.

Levey, A. and Lindenbaum, M., 2000, 'Sequential Karhunen-Loeve basis extraction and its application to images', *IEEE Transactions on Image processing*, vol. 9, no. 8, pp. 1371–1374.

Ma, C., Huang, J.-B., Yang, X. and Yang, M.-H., 2015, 'Hierarchical convolutional features for visual tracking', in *Proceedings of the IEEE International Conference on Computer Vision*, pp. 3074–3082.

Ning, J., Yang, J., Jiang, S., Zhang, L. and Yang, M.-H., 2016, 'Object tracking via dual linear structured svm and explicit feature map', in *Proceedings of the IEEE Conference on Computer Vision and Pattern Recognition*, pp. 4266–4274.

Platt, J., 1998, 'Sequential minimal optimization: a fast algorithm for training support vector machines', *Technical Report*.

Qi, Z., Ting, R., Husheng, F. and Jinlin, Z., 2012, 'Particle filter object tracking based on Harris-SIFT feature matching', *Procedia Engineering*, vol. 29, pp. 924–929.

Ross, D. A., Lim, J., Lin, R.-S. and Yang, M.-H., 2008, 'Incremental learning for robust visual tracking', *International Journal of Computer Vision*, vol. 77, no. 1–3, pp. 125–141.

Sun, L. and Liu, G., 2011, 'Visual object tracking based on combination of local description and global representation', *IEEE Transactions on Circuits and Systems for Video Technology*, vol. 21, no. 4, pp. 408–420.

Wang, D., Lu, H. and Yang, M.-H., 2013, 'Online object tracking with sparse prototypes', *IEEE Transactions on Image Processing*, vol. 22, no. 1, pp. 314–325.

Wang, Z., Crammer, K. and Vucetic, S., 2012, 'Breaking the curse of kernelization: budgeted stochastic gradient descent for large-scale svm training', *Journal of Machine Learning Research*, vol. 13, pp. 3103–3131.

Wu, Y., Lim, J. and Yang, M.-H., 2013, 'Online object tracking: a benchmark', in *Computer vision and pattern recognition (CVPR)*, pp. 2411–2418.

Zhang, J., Ma, S. and Sclaroff, S., 2014, 'Meem: robust tracking via multiple experts using entropy minimization', in European conference on computer vision, pp. 188–203.

Zhang K. and Song, H., 2013, 'Real-time visual tracking via online weighted multiple instance learning', *Pattern Recognition*, vol. 46, no. 1, pp. 397–411.

Zhang, K., Zhang, L., Liu, Q., Zhang, D. and Yang, M.-H., 2014, 'Fast visual tracking via dense spatiotemporal context learning', in European conference on computer vision, pp. 127–141.

Zhang, K., Zhang, L. and Yang, M.-H., 2013, 'Real-time object tracking via online discriminative feature selection', *IEEE Transactions on Image Processing*, vol. 22, no. 12, pp. 4664–4677.

Zhou, H., Yuan, Y. and Shi, C., 2009, 'Object tracking using SIFT features and mean shift', *Computer Vision and Image Understanding*, vol. 113, no. 3, pp. 345–352.

11

Feature, Technology, Application, and Challenges of Internet of Things

Ayush Kumar Agrawal and Manisha Bharti

CONTENTS

11.1 Introduction

Internet of Things (IoT) deals with device-to-device communication and also with interconnected connections where all IoT-embedded devices are linked to every other, which includes automotive devices and self-learning devices. IoT is associated with a major part of the internet and named as an important international network framework with self-setup capabilities. It relies on normal and practical communication protocols wherever physical and virtual "things" have identities, physical attributes, and virtual personalities that use intelligent interfaces, and are seamlessly integrated into the knowledge network. Typically, this may be explicit because of the interconnection of several things/devices.

In this chapter, the authors describe the options and technologies that are employed in the Web of Things and can be used in day-to-day life. Varied options embrace unequivocally acknowledgeable and available objects like computing, designing, and geo-localization and Size concerns. Technologies like RFID, local area network IEEE 802.11, QR Code, ZigBee IEEE 802.15.4, sensors and smartphones are taken into thought. Today, the current generation is leaning towards sole tech, and, currently, technology is the "Internet of things." Many other uses of IoT are outlined in this chapter, including traffic observation, health, security, transport and supplies, lifestyle, and demotics.

The impact of the internet has modified the lifetime of an individual with a revolution is a very important issue to be celebrated before the exploitation of any technology for any purpose. As the Internet of Things is associated with forthcoming technology, there will be societal issues to overcome relating to individuals, security, privacy, environmental aspects, and technological problems; these are important issues to be taken care of before the use of any technology.

11.2 About the Web of Things

To have a complete discussion of the Web of Things is the main challenge because these technologies have not only one sort of hardware but several sorts of devices that are not linked but that existed. Good devices, including linked thermostats or internet-accessible "Smart ovens," are examples that independently started to point at the surface of the IoT service potentialities. Many people think about changing towards IoT, where each device linked will have a separate IP address. The IoT history changes on to counsel as this IoT can cover various devices coupled with the web, relating data concerning surroundings, logistics, and management systems. Some counsel the IoT are linked as a part of a bigger interconnected infrastructure that will automatically collect and manufacture knowledge concerning the surroundings. IoT devices aren't PCs or any other type of wireless telephony devices.

One of the most trusted suppliers of tech merchandise files, as reported, is Gartner, for its research of the "Gartner Magic Quadrant" analysis method; as an example for our functions it is useful, loosely describing IoT. In line with Gartner merchandise promulgation upon this subject, "The Internet of Things is the network of physical objects that contain embedded technology to speak and sense or interact with their internal states or the external surroundings". Including the economic value encompassing the IoT, "IoT product and repair suppliers can generate progressive revenue prodigious $300 billion, principally in services, in 2020. It'll lead to $1.9 trillion in international economic value-add through sales into various finish markets."

For information functions generally, and also the specific IoT information-involved report comprised herein, we are going to use an operating explanation of IoT. As in early days, IoT essentially do not have the technologies which we are looking at now, and our complete study can be attached with that of IoT networks—commonly known by phase that shows a Bluetooth signal that facilitates to supply the main awareness of the location as per the latest modern use.

The main problem as seen by researchers is fascinating, as it describes important and helpful IoT uses—which come back unbound. In Rose's perspective, it might be to think about the wireless telephony-enhanced IoT surroundings as a transformation state encompassing IoT during which devices are less bound to our management systems of wireless telephony devices, personal computers, or servers and act just like the "enchanted objects" delineated in his IoT vision.

Therefore, the reason IoT technologies are hyped nowadays is their promise for a greater interconnected globe, as well as the expected benefits that these deeper connections can deliver. The IoT was noted by at least one research author as "the network of networks" and "the original true evolution of the planet's broad network," and in some sectors, it is capable of providing even more automation than is currently feasible with non-linked infrastructure.

The delay here to imagine a deeply connected world is not new. Leading the North American nation to ask: what is the purpose of departure using information robotics and wireless telephony information? There is evidence to advise that the IoT consists of little laptop assets pervasively connected to the world and directly linked to the cloud. If projections are evidently playing out, these extremely small, omnipresent-networked devices on the far-sided implementation of computers, laptops, or wireless telephony can improve in number. IoT is the fruit of numerous computing forces (such as wireless telephony techniques), demonstrated in the unfolding of the smartphone, the processing of cloud-based infrastructure data streams, and also the ever-smaller kinds of networked computing components.

11.3 IoT as Tool for Change in Technology

Telecommunications is presently one amongst the quickest dynamical industries with broadband networks and repair suppliers sharply competitive in their advanced spread subscription points and additional facilities. The change from voice to content packages in streaming, video, HDTV, etc. occurred too quickly for merchandisers to handle the demand. A large part of this can be ascribed to the wide variety of excellent wireless handheld telephony devices used in computing.

The convergence of standards/innovations and talent for attaching smaller devices, items, and sensors affordably and simply produced a hyper-linked universe that bridges the virtual and technical to get, process, swap, and consume information for the IoT web. The "Things Web" may be a union of world norms, techniques, and devices that zone units capable of communicating within the virtual environment. This type of technology is used to generate process, exchange information, and make decisions. With its multiple consequences and large device proliferation, IoT is commonly believed to be one of the greatest revolutions in the new millennium. Its outcome has the ability to be felt in all industries and professions on a global scale.

Several corporations favor their use and services with entirely distinct categories of IoT platforms. Product and service manufacturers use their IoT as "Industrial IoT," while others use corroborated IoT devices such as wearable devices or locations such as "Smart Home" and "Smart City." IoT devices are not substitutes for people, but they overcome individual limits. Using drones with cameras and sensors can make the journey to places where individuals cannot reach and immediately organize, store, and send information to a sensitive device. As device makers are creating IoT devices for remote and simulation-based observation, there are intrinsic difficulties such as moment to plug, capacity, authentication, safety, preservation of digital information, and overcoming technical issues such as power consumption, and limited computing technique. With cheap sensors and new sponsoring technologies such as the 4G LTE Cat-M1 system, IoT adds to nearly every industry an increase in potency and performance.

11.4 Characteristics of IoT

The Internet of Things ensures a lot of use in human lives, making life simpler, safer, and more sensitive. Many units use sensitive towns, housing, transportation, energy, and sensitive settings.

The IoT's important properties are as follows:

11.4.1 Heterogeneity

The IoT systems are autonomous, looking on varied platforms and channels of hardware. They will communicate through separate networks or channels with alternative appliances or service platforms.

11.4.2 Interconnectivity

Globally, using IoT, one thing is interconnected with the infrastructure of information and communication. IoT is able to generate connected services at intervals; the limitations of things include private security and linguistic cohesion between physical things and their related virtual things. Among the scheme for delivering services at intervals, the limitations of things will change every technology in the physical globe and in the data world.

11.4.3 Dynamic Changes

As endorsed entirely separate hardware operating systems and channels, the devices among the IoT zone unit freelance. They will move through entirely separate networks or channels with appropriate devices or service operating systems.

11.4.4 Enormous Scale

The number of devices that need to be maintained that interact with each other is at least an associated order of magnitude larger than the devices combined to the present net. Additionally, it is essential to manage the produced information and interpret it for execution purposes and to link it to data linguistics as an efficient scientific discipline.

11.4.5 Safety

We cannot ditch security; we tend to benefit from the IoT. We need to style for safety as both IoT maker and beneficiary. This includes our non-public knowledge security and our practical well-being security. Secured endpoints, networks, and moving data across it all implies establishing scale-up safety.

11.4.6 Connectivity

Connectivity permits accessibility and compatibility to the network from one end to another. Accessibility on a network is accepted whereas compatibility offers the current capability to get and consume data.

11.5 Applications of IoT

We have covered the notion of useful objects which can cover networked devices. Interconnected devices are promising to provide additional practicality, comfort, and generally duplicated living standards. Take the icebox that will order a preset food type once it goes below a preset threshold. The IoT will cover a lot of home equipment like this, enabling some to explain the IoT because as everything's net. As an associate instance of the likelihood of associated linkage explosion, consider the examples outlined in the following sections.

11.5.1 Smarter Cities

The demands of good cities involve careful designing,, with the convenience of government and voters agreeing to adopt the Web of Things technology.

With IoT, cities are often created at several levels, enhancing infrastructure, stepping up public transportation, reducing traffic blockage, maintaining individual safety, health, and community. By connecting all town services like health systems, transport systems, weather observation systems, and so on, net users everywhere can access information on services operating below protocols, and cities can become intelligent through the Web of Things.

11.5.2 Smarter Home

In-house self-regulation techniques for Wi-Fi were used primarily thanks to the network nature of used natural philosophy wherever Wi-Fi typically supports equipment like wireless telephony devices, TVs, etc. Wi-Fi has begun to become a part of the house IP network, and wireless telephony computing devices like smartphones, tablets, etc. are progressively being adopted, for example, networking to supply network or on-line streaming services to manage the practicality of the device over the network. At the same time, wireless telephony devices offer customers access to a portable controlle' for network-linked natural philosophy. Both tool types are often used as IoT gateways. Several professions are considering developing platforms that incorporate construction mechanization with recreation, power following, tending following, and wireless sensing element following in home and building settings. Good lighting, intelligent surroundings and media, air management and heating, energy management and safety are among the foremost necessary applications of IoT in intelligent homes and buildings through the notion of the Web of Things, homes, and buildings. In addition to the apparent monetary and environmental advantages, wireless sensing element networks (WSNs) with integration of tech into the web can offer good energy management in homes. In conjunction with power management schemes, the web provides a chance to access the energy knowledge and management systems of a building from a laptop computer or smartphone from anywhere on the globe. The longer term Web of Things can cause a sensible building management system, which will be thought of as a part of a far larger data system utilized by construction facility managers, to manage the utilization of electricity and power..

11.5.3 Smart Energy

Data and management are linked to a sensible grid and have created intelligent energy management. A sensible grid that integrates knowledge and data technologies into the ability network can give two-way period communication between suppliers and customers, generating additional spirited energy flow interaction that may assist sustainable and efficient energy production. The key elements of ICT can mix sensing and following systems for energy flows; digital communications infrastructure are wont to send knowledge across the grid; in-home meters for monitoring energy use; coordination, management, and automation systems for aggregating and processing information; and very interactive, responsive electricity generation. Thanks to the Web of Things for intelligent grids (like industrial, solar, and atomic energy, as well as cars, hospitals, and cities) several applications are often self-addressed, permitting the Web of Things essentially within the good grid facet.

11.5.4 Smart Health

Close attention is required for the utilization of IoT police work techniques for hospitalized patients whose physiological standing ought to be monitored frequently. Sensors are used for intelligent health to get in-depth physiological knowledge and to use gateways and the cloud to look at and retrieve data, then wirelessly transmit the analyzed knowledge to caregivers. It replaces the method of obtaining a healthcare worker to verify the important signs of the patient at frequent intervals, instead providing a continuing machine-driven stream of information. It will increase the standard of care through periodic care and lower the cost of care by reducing the cost with respect to data assortment and analysis. Several people around the world are plagued by dangerous health because they do not have instant access to economical police work of health.

11.5.5 Environmental Observation (Smart Appliances)

There are many retailer products that demonstrate IoT technology, like the wireless charger. These appliances will learn the practices, and heating and cooling habits, of a home's residents over time by using user input of temperature preferences in order to optimize the preferred temperature. Often, nest thermostats are paired and optimized without the wireless telephony device of a user. A wireless telephony app offers some capacity for householders to handle and collect ideas about their home environments.

11.5.6 Smart Vesture and Good Accessories (Wearable)

For observation pulse, power per unit region, and comparable information, medical information will be transmitted to a central server. The info could provide health observation prognostication analytics. In a piece from Cisco's Chief Futurist, it was totally observed that these capacities can develop profoundly over "the next few years, we'll be able to swallow a pill that will monitor our gastrointestinal tract and showing intelligence send relevant data to our doctors at the correct time and within the context of what we're doing. Expectant mothers can wear 'smart tattoos' to observe the health and activity of their babies, associated send their doctor an early alert once labor begins. We've solely begun to scratch the surface of however wearable tech can remodel our lives."

11.5.7 Hobbyists

A number of hobbyists are also fascinated by the IoT region. We will draw on particular programmable microcontroller board examples after removing many of the techniques that inherit the IoT in Chapter 3. However, these small computers have the option of owning extra peripherals such as storage and displays. Their region unites a range of possibilities within the IoT space for hobbyists. In some respects, owing in part to the pliability of the devices, these have a number of the most attentive applications.

11.6 Challenges

Within the price of implementation, there are some challenges to applying the concept of the Web of Things, for example, the expectation that the tech should, with massive number of objects, be accessible at low value. IoT faces several other difficulties, too.

11.6.1 Scalability

The Internet of Things incorporates a bigger plan than the quality laptop internet because things work along in associated open settings. Therefore, basic practicality like the discovery of communication and repair should work equally effectively in both small- and large-scale settings. To realize an efficient quantifiability operation, the IoT desires contemporary options and techniques.

11.6.2 Self-Organizing

Smart things must not be managed as PCs, requiring the configuration and adaptation of their users to specific circumstances. Wireless telephony things, usually used solely periodically, have to be compelled to connect *ad libitum* and be able to organize and put themselves together to suit their specific surroundings.

11.6.3 Data Volumes

Some Internet of Things implementation situations would force rare communication and aggregation. Large-scale networks can collect huge quantities of data on central network nodes or servers. The term that reflects this development is *massive knowledge*, which, in respect to contemporary techniques for storage, process, and management, needs several operational processes.

11.6.4 Data Interpretation

To help users, it is necessary to interpret as well as possible the native context determined by sensors. To change service suppliers to require advantage of the disparate data that may be made, the taken sensing element data should be ready to draw some generalized conclusions.

11.6.5 Interoperability

Every style of the intelligent object on the item's net has distinct capacities for knowledge, process, and interaction. Completely different objects would even be subject to variable circumstances like the provision of energy and the needs for communications information measure. Common standards are required to push the communication and collaboration of those objects.

11.6.6 Automatic Discovery

Suitable services for things should be mechanically recognized in dynamic settings that desire adequate linguistics to explain their practicality.

11.6.7 Software Complexity

In order to control the good objects and supply services to help them, a bigger computer code infrastructure is required on the network and on background servers. That is because, as in normal embedded systems, the computer code systems in good objects can operate with token resources.

11.6.8 Security and Privacy

Other demands in Web of Things would be essential in respect to the protection and safety parts of the web, such as the confidentiality of communications, the believability and the trait of communication partners, and the integrity of messages. There's a desire to access bound facilities or avoid communication with alternative things in IoT, and company operations involving intelligent things ought to be secure from prying eyes of rivals.

11.6.9 Wireless Communications

From an influence perspective, well-established wireless systems like GSM, UMTS, Wi-Fi, and Bluetooth are way less acceptable; the latest WPAN standards like ZigBee et al., still below growth, could have a smaller information measure; however, they use significantly less power.

11.7 The Problem of Overlays

You might suppose Wi-Fi would be a natural alternative for enterprise IoT use. After all, most enterprises already use Wi-Fi infrastructure. For a range of reasons, though (power constraints, a desire for smaller kind factors, a conflict between IoT knowledge needs and high-capacity Wi-Fi connections), Wi-Fi is commonly not the most effective radio choice for IoT.

Instead, the devices at the core of a number of the fastest-growing enterprise IoT trends (smart door locks, wearable workers alert devices) use IoT-optimized technologies like LoRa, BLE or ZigBee.

11.7.1 Redundant Overlay Networks

The proportion profit will be compelled to justify building a replacement, separate network, with separate wiring and power associated security.

11.7.2 Management Complexity

Managing those redundant networks is additionally painful. Several product classes that will not be managed singly (routing and shift, wireless and security) are converging and, consequently, may spur the merchandise toward consolidation. As a network merchandiser, if you cannot bring multiple systems below one management, several IT organizations will not work with you.

11.7.3 Precious Physical Space

Notwithstanding enterprises are willing to think about overlay networks, several merely do not have or will not commit the physical property to deal with additional network elements. Every new IoT tech needs its own entry and dedicated firewall, and separate shift, powering and cable infrastructure; enterprises would need somewhere to place all that equipment.

Together, these problems produce a vital barrier to wide-scale adoption of enterprise IoT technologies—even once the technologies themselves will yield significant merchandise advantages. As a result, outside of sectors like industrial, the sole enterprises deploying IoT nowadays tend to be those with the most to gain or the most to lose.

Take cordial reception, one amongst the few enterprise sectors seeing major IoT investments (in this case, in workers alert systems). While not recent, negative press around harassment of cordial reception employees, and also the new rules and merchandise initiatives in response, several of those would still seemingly be stuck within the strategy planning stage.

11.8 Rise of Converged APs

Fortunately, the complexities and inefficiencies of the established order do not get to persist forever. There is a path forward for enterprise IoT: convergence. Indeed, as enterprises look to try to add with less and keep up with tech modification, they are rigorous converging on several fronts, from unified wired and wireless management to useful knowledge center platforms. In these surroundings, the conventionally "commoditized" AP is turning into a hotbed of innovation.

In some ways, this can be a continuation of long trends, like APs that support multiple Wi-Fi radio bands. The newest APs push convergence even farther. New technologies just like the tumult R730, as an example, mix dual-band Wi-Fi with embedded BLE and ZigBee radios. Platforms like these can enable enterprises to combine and match radio modules for alternative IoT radio technologies too—all running below a similar enclosure. All of a sudden, IoT technologies that were antecedently too complicated and expensive become viable for nearly any enterprise.

11.9 Addressing Convergence Challenges

Of course, convergence will bring its own problems. Every wireless radio tech incorporated into a conventional "Wi-Fi only" AP will add new technological challenges.

11.9.1 Radiofrequency (RF) Interference

One issue several fashionable IoT radio standards have in common is that they operate over the two. Gig cycle spectrum—one already thronged with Wi-Fi traffic and interference from microwave ovens, conductor phones, etc. Implementing multiple radios sharing a similar spectrum, among a similar enclosure, while not decreasing performance is not easy.

11.9.2 Packet Coordination

Running multiple distinct wireless applications over a similar infrastructure would force additional refined traffic handling.

11.9.3 Antenna Design

To boost connections and minimize interference, antennas for the various radios (again, synchronal among a similar enclosure) have to be compelled to be designed with suitable spatial separation.

If converged APs are progressing to break the enterprise IoT logjam, they have to handle these and alternative tech problems. That needs skillful RF and antenna engineering, and complicated AP styles engineered to fulfill the various desires of recent enterprise use cases. Not each merchandiser has that level of experience.

Ultimately though, converged APs play a giant role in lowering the barriers to enterprise IoT adoption. As additional enterprises have the benefit of IoT apps, they will even notice exploitation converged APs to be not simply more cost-effective, but also simpler, period. After all, even once exploitation separate overlay networks for multiple IoT technologies, the radios still operate within the same airspace. Therefore, they're still subject to a similar RF interference and linked problems. Deploying wireless technologies engineered from the bottom up to use multiple radios, rather than making an attempt to resolve those issues can progressively make sound merchandise sense.

11.10 Future Technologies of IoT

The IoT is connected with increasing space, and because of an increasingly interlinked networked environment, several other feasible services related innovations could be obtained. However, there is speculation about its diverse manifestations that can affect our lives and the services that we can deliver in and out of library doors. One theorist argued that the IoT had been implemented, "a good planet can evolve, wherever several of the everyday things around the North American nation have an associated identity in Net, acquire intelligence, and mash-up data from various sources." However, the computer code components needed to generate this have not been created since most hardware-based IoT solutions are not integrated into intelligence-collection networks. Kopetz jointly observed "the IoT's novelty is not in any fresh riotous tech, but in the omnipresent readiness of fine objects," so it should not be just one effect from one IoT tech application. Instead, the IoT stands to be an accumulative tech result because of its pervasive nature.

The theoretical and alleged IoT benefits for libraries involve issues, however, researchers are able to blend information that can be produced or consumed from IoT devices to deliver innovations in commission understanding, which in reality can lead to greater automation. In fact, information that is produced by inner library control may promote greater perception by assortment designers, but users behave with physical fields. Before the IoT, with relevance to the evaluation of the physical information region, there was no reasonable outfit to know; user engagement appeared to be at a pervasive stage in collections and control points. On the evaluation of the far side, a greater insight into the specific use of the data region may allow libraries to raise awareness of area utilization stories and create evidence-based selections.

The demand in education for evidence-based deciding has never been stronger. Whereas there has been a lot of study by ethnographic researchers, WHO collect qualitative knowledge concerning what students would really like to try in areas; a deep understanding entails real quantitative use knowledge concerning information areas. An associate actively financed Knight Foundation Project, which uses IoT technologies to support the evaluation of fields. The Live Long-Term Project aims to produce hardware and computer code alternatives that can provide your data building with a "Google-Analytics-style dashboard," the range of visits, what patrons browsed, what elements of the information were busy throughout those elements of the day, and more. In the long term, this will happen by exploiting easy and cheap sensors, which will collect knowledge concerning building usage that's currently invisible. Creating these invisible occurrences can enable librarians to create strategic selections that make additional economical and effective experiences for his or her patrons.

11.10.1 Cloud Computing

Cloud and IoT's are two important words that will lead to autonomous development. These worlds are terribly distinct from one another; however, their options are usually complementary, particularly wherever IoT will cash in on the cloud's nearly infinite capacities and resources to offset its technological limitations like storage, processing, and communication. The cloud will offer an economical answer for managing and composing IoT services, implementing use and services that exploit the things or data that they manufacture. On the opposite side, by increasing its reach to cope with real-world things in distributed and dynamic manner, and by providing contemporary services in a massive quantity of real-life things, the cloud will exploit IoT. In several instances, the cloud will provide the common layer between things and use, protecting all the complexity and practicality needed to implement the latter. This can affect future apps wherever collecting, processing, and information systems can create modern problems, especially in multi-cloud or fog cloud conditions. The cloud supports IoT apps to change data assortment and data processing, to expand quick set-up and inclusion of latest things; while maintaining low reading and complicated scientific discipline expenses, the cloud is the most acceptable and efficient answer to handle data generated by IoT, generating contemporary potentialities for data assortment, integration, and sharing with third parties. Once within the cloud, data are often handled as unvaried through robust-defined Apis, protected through the applying of high-level safety, and may be accessed and viewed from anywhere directly.

11.10.2 Shared Computing

Shared computing uses networked laptop teams for a comparable computer purpose; shared computing has many present problems with synchronous and parallel computing, as all three fall within the science computing industry. Cloud computing has begun nowadays with a giant quantity of shared computing techniques combined with virtualization of hardware, service-oriented design, and autonomous and utility computing. The mutual computation Internet of Things represents a vision wherein the internet embodies everyday things into the globe of reality. Physical things are no longer unlinked from the virtual surroundings; however, they will be managed remotely and may act as physical net services' access points.

11.10.3 Cloud Computing

Cloud computing recalls the edge computing of the cloud. Not like the cloud, fog platforms were defined as small machine technologies at the historically low levels of the network. Such platform options allegedly have low latency, location awareness, and management of wireless access. Whereas edge computing or edge analytics specifically refers to activity analytics on or near the network's purpose, a fog computing system would offer analytics from the heart of the network. IoT can be implemented by means of fog computing, wherever computing, storage, management, and networking power is maintained in the design of data centers, the cloud, top devices such as gates or routers, or edge devices such as PCs or sensors themselves.

11.10.4 Wireless Fidelity (Wi-Fi)

Wireless Fidelity (Wi-Fi) may be a networking tech that allows PCs and alternative devices to operate via a wireless signal. Vic Hayes has been named the male parent of Wireless Fidelity. NCR Corporation created the precursor to Wi-Fi in the Netherland in 1991. The primary wireless product was launched on the market with speeds from one Mbps to two Mbps below the name Wave computer network. Today, nearly ubiquitous Wi-Fi provides high-speed Wireless native space Network (WLAN) to various offices, households, and other places like restaurants, coffee shops, airports, etc. Integrating Wi-Fi into notebooks, handhelds, and shopper physical philosophy devices have accelerated Wi-Fi acceptance wherever these phones are almost default. Technology includes any type of dual-band IEEE 802.11, 802.11a, 802.11b, 802.11 g, and 802.11n product support. Today, the units of Wi-Fi corridors are being moved via wireless APs.

11.10.5 Bluetooth

Bluetooth wireless tech is a low-cost, narrow-range radio tech that removes the need for proprietary wiring between devices such as notebook PCs, hand-held PCs, PDAs, cameras, and printers, within an economical 10–100 meter variety. In addition, it usually operates on but one Mbps, and Bluetooth uses the ordinary specification of IEEE 802.15.1. Ericson launched a project known as "Bluetooth" in 1994. It is used for private area network (PAN) development. Pico net could be an assortment of Bluetooth appliances that love current communicating. At a knowledge sharing point, this Pico net is capable of 2–8 computers, and information is often text, video, picture, and sound. The Bluetooth interest group cluster involves as many as one thousand professions with Intel, Cisco, HP, Aruba, Intel, Ericson, IBM, Motorola, and Toshiba.

11.10.6 ZigBee

ZigBee is one of the protocols created to reinforce wireless sensing element network characteristics. The 2001-based ZigBee Alliance develops ZigBee tech. ZigBee's features are low price, low rate, relatively transient vary of transmission, quantifiability, dependableness, versatile protocol style, and it's associate IEEE 802.15.4-based low-power wireless network protocol. ZigBee incorporates a form of concerning one hundred meters and a 250 kbps information measure, and the topologies it operates as a star, cluster tree, and mesh. It's widely employed in industrial controls, home automation, digital farming, medical observation, and energy systems.

11.11 Cloud Computing in IoT

11.11.1 Remote Process Power

The Internet of Things will not simply relegate itself to the "intelligent fridges" that come to mind when someone mentions this new wave of technical school. On the contrary, ultimately everything can be a sensitive device, swinging fresh requirements on the strength of the raw method. In addition, with miniaturization growing and 4G property becoming even more prevalent, the cloud can return to the rescue, allowing developers to dump cloud-computing services to process.

11.11.2 Lowers the Entry Bar for Suppliers Who Lack the Infrastructure

The IT revolution was only possible due to the steady lowering of the entry bar for innovators and developers, leading to a world where a good plan succeeds without the support of an oversized company. Remember a number of the most significant names and trends in technical college—the credit for many of them goes to young, skilled, creative vicissitudes of WHO. The cloud can make it simpler for such innovators to hit the IoT revolution, offering their devices and services with a ready-made infrastructure.

11.11.3 Analytics and Observation

An extra seamless knowledge could be one of IoT's core tenets. In this respect, the cloud can promote access to sophisticated analytics and observation by device producers and repair providers. It is a move towards continual innovation for the developer, while for the customer it holds the promise of greater reliability and extra seamless knowledge.

11.11.4 User Security and Privacy

The effect it has had on our privacy and safety has been significant concern about the quality revolution. Recent devices and facilities are generally lagging behind, putting consumers understanding and privacy at risk from security exploits that have lately been uncovered. Despite touching every facet of our life, the IoT could actually sidestep this safety problem, all because of the cloud. With devices that use popular Apis and back-end infrastructure, major safety updates can be instantaneous and blanket.

11.12 Challenges in Integration of Cloud Computing and IoT

However, the convergence of the cloud and IoT have been stated, providing many benefits and encouraging the birth or enhancement of the type of attention-grabbing use. At a comparable moment, we saw that the complicated situation of inclusion imposes many difficulties for each implementation that the analytical community is currently getting attention.

11.12.1 No Uniformity

The broad non-uniformity of devices, operating systems, platforms, and services that can be obtained and probably used for brand new or enhanced applications is a major challenge. Non-uniformity cloud platforms are also a non-negligible problem. Cloud services usually accompany proprietary interfaces, inflicting the accurate provider's resource integration and mashup to be correctly customized. This problem is often exacerbated once customers adopt multi-cloud methods, i.e., once services depend on various providers to increase application performance. Only partially solved by cloud brokering, these aspects are willingly implemented by cloud providers or third parties. In general, IoT facilities and use were scheduled as isolated vertical alternatives, during which all the system components were closely linked to the accurate implementation context. Suppliers have to survey target scenarios, analyze needs, select hardware and computer code environments, integrate heterogeneous subsystems, create and deliver computer infrastructure, and deliver service maintenance for every feasible application/service.

11.12.2 Performance

In particular, achieving stable acceptable network performance in order to succeed in the cloud could be a major challenge, given that the rise in broadband did not follow the evolution of storage and computation. In fact, the provision of information and services must be carried out with high reactivity in many situations (e.g., once quality is required). Poor QoS can also be littered with usability and user knowledge (e.g., once multimedia system streaming is required).

11.12.3 Dependableness

When cloud-IoT is adopted for mission-critical use, reliability problems generally occur. For example, in the context of excellent performance, moving vehicles, as well as transmission networking and communication, is generally intermittent or unreliable. Various difficulties connected with device failure or non-invariably available devices occur once applications were deployed in resource-affected settings. Cloud capacities, on the one side, make it easier to overcome a number of these difficulties (e.g., the cloud enhances the device's reliability by allowing important tasks to be dumped and thus extending the battery length of devices or creating the probability of constructing a modular architecture); on the other side, it presents uncertainties associated with the virtualization or resource center of information. The lack of reliability analyses and case studies exacerbate the challenge.

11.12.4 Massive Scale

Cloud-IoT allows new application styling targeted at integrating and analyzing information from (integrated) universal devices. Many of the delineated circumstances implicitly require communication with a very large number of such devices, typically spread across wide-area settings. Typical difficulties are made more durable by the huge scale of the following schemes. In addition, IoT device allocation makes observation tasks more durable as they need to deal with the dynamics of latency and ownership issues.

11.12.5 Big Data

With the associated calculable frame of fifty billion devices that can be networked in the near future, special attention should be paid to transporting, storing, accessing, and processing the large amount of information that they will produce. IoT is one of the biggest sources of information owing to the latest technological innovations, and the cloud can alter to store it for a prolonged period and conduct complex analyses on it. The existence of wireless telephony devices and the generality of sensing elements is thus concerned with ascending computing platforms (2.5 big bytes of data are produced each day). Handling this knowledge handily could be a vital challenge because the overall application performance is extremely addicted to the properties of the information management service. Sadly, no excellent knowledge management answer exists for the cloud to manage massive knowledge.

11.13 Conclusion

In this chapter, the author wants to make us understand that the Internet of Things is now in the first steps of evolution or development and have a lot of potential. It looks like a bias to have one of these products, and that is good for the future of the Internet of Things. As we understand it in this chapter, the prognostications of many people are very useful for this sector; using this type of product supports a lot of gains, production, time, and so on. The attraction is huge for many companies to shift into the utilization of those products. As we have seen, people who are utilizing those type of products cut the costs. It is sure that those products are probably expensive, but they offer a good return on investment. Internet of Things is a new technology which presents many applications to connect things to things and human to things over the internet. Each object in the world can be identified, connected to each other through internet taking decisions independently. All networks and technologies of communication are managed in building the concept of the internet of things; such technologies are RFID, mobile computing, wireless sensors networks, and embedded systems, in addition to many algorithms and methodologies to get management processes, storing data, and security issues. Internet of Things requires a standardized approach for architectures, protocols, and identification schemes, and frequencies will happen parallels, each one targeted for a particular and specific use. Through the Internet of Things, many smart applications become real in our life, which enables us to reach and contact everything, in addition to facilities, many important aspects for human life such as smart healthcare, smart homes, smart energy, smart cities, and smart environments. The author tried to make us familiar with the Internet of Things system that our devices use. As we tend to conclude this chapter, numerous factors are noted. The information is gathered to help promote what is known as machine learning in most instances.

Machine learning can be an AI method that helps machines "learn" while not being programmed by someone. During a strategy that focuses on the information they obtain, the computers are programmed. This fresh understanding will then make it easier for the machine to "learn" your preferences and subsequently change itself. For instance, once a video page indicates something you may want, your preferences may have been learned from your previous choices.

Major wireless carriers will keep rolling out the 5G network. 5G-cellular wireless networks promise higher speed and can connect more smart devices simultaneously. Faster networking means collecting, analyzing, and managing the information accumulated by your smart devices to a large degree. This will stimulate innovation in businesses producing IoT equipment and boost customer demand for fresh products. The auto industry will shift to high gear with the introduction of 5G. The growth of driverless cars-as well as on-road vehicles-will benefit from quicker data movement.

You cannot see your vehicle as a computer for the Internet of Things. Nevertheless, new cars will evaluate your information quickly and connect to other IoT appliances including other four-wheel high-tech vehicles. In time, compared to Wi-Fi routers, 5G IoT devices will connect directly to the 5G network. This trend will render these phones more susceptible to direct assault, according to a recent blog post from Symantec.

It will become harder for home users to monitor all IoT equipment as they bypass the main router. Increasing dependence on widespread, cloud-based storage will provide fresh objectives for attackers to try a violation. Attacks on botnet-driven denial of distributed service (DDoS) have used infected IoT equipment to downgrade websites. IoT systems can be used to guide other assaults, according to the blog post from Symantec. There may be future efforts to arm IoT equipment, for instance. The shutdown of national thermostats in an enemy state during a harsh winter would be a rough illustration.

IoT device development is just one reason to raise issues about safety and privacy. The European Union in mid-2018 introduced the General Data Protection Regulation. In many nations around the globe, the GDPR has taken comparable safety and privacy measures. California has lately enacted a hard law on privacy in the United States.

References

Al Nuaimi, E., Al Neyadi, H., Mohamed, N. and Al-Jaroodi, J., 2015, 'Applications of big data to smart cities', *Journal of Internet Services and Applications*, vol. 6, no. 1, p. 25.

Bhushan, D. and Agrawal, R., 2020, 'Security challenges for designing wearable and IoT solutions', in *A handbook of internet of things in biomedical and cyber physical system*, Springer, Cham, pp. 109–138.

Ezechina, M. A., Okwara, K. K. and Ugboaja, C. A. U., 2015, 'The Internet of Things (Iot): a scalable approach to connecting everything', *The International Journal of Engineering and Science*, vol. 4, no. 1, pp. 09–12.

Fatima, F., Husain, B. N., Azharuddin, M. and Mabood, M. A., 2015, 'Internet of things: a survey on architecture, applications, security, enabling technologies, advantages & disadvantages', *International Journal of Advanced Research in Computer and Communication Engineering*, vol. 4, no. 12, pp. 498–504.

Fisher, D., 2016, 'Internet of Things: Converging technologies for smart environments and integrated ecosystems', *DocPlayer*. Available at: http://docplayer.net/1073 234-Internet-of-things-converging-technologies-for-smart-environments-and-i ntegrated-ecosystems.html.

Gubbi, J., Buyya, R., Marusic, S. and Palaniswami, M., 2013, 'Internet of Things (IoT): a vision, architectural elements, and future directions', *Future Generation Computer Systems*, vol. 29, no. 7, pp. 1645–1660.

Haav, H. M., 2014, 'Linked data connections with emerging information technologies: a survey', *IJCSA*, vol. 11, no. 3, pp. 21–44.

http://cdn2.hubspot.net/hubfs/552232/Downloads/Partner_program/Smart_E nvironmntsFlyer.pdf?t=1458917278396.

7wData, 2015, 'How Big Data and Internet of Things builds Smart Cities', *7wData*. Available at: http://www.7wdata.be/article-general/how-big-data-and-intern et-of-things-builds-smart-cities/.

Kaur, S. and Singh, I., 2016, 'A survey report on internet of things applications', *International Journal of Computer Science Trends and Technology*, vol. 4, no. 2, pp. 330–335.

Kumar, A., Kumar, P. S. and Agarwal, R., 2019, March, 'A face recognition method in the IoT for security appliances in smart homes, offices and cities', in *2019 3rd international conference on computing methodologies and communication* (ICCMC), IEEE, pp. 964–968.

MeraEvents, 2016, Workshop on Internet of Things (IoT) Using Raspberrypie – 2 Days, *MereEvents*. Available at: http://www.meraevents.com/event/iot-workshop.

Misra, S., Maheswaran, M. and Hashmi, S., 2017, *Security challenges and approaches in internet of things*, Springer International Publishing, Cham.

Murray, A., Minevich, M. and Abdoullaev, A., 2011, 'Being smart about smart cities', *Searcher*, vol. 19, no. 8., pp. 331–340.

Odulaja, G. O., Oludele, A. and Shade, K., 2015, 'Security issues in internet of the things', *Computing, Information Systems, Development Informatics & Allied Research Journal*, vol. 6, no. 1, pp. 33–40.

Preethi, D. N., 2014, 'Performance evaluation of IoT result for machine learning', *Transactions on Engineering and Sciences*, vol. 2, no. 11.

Raza, S., 2013, *Lightweight security solutions for the internet of things*, Doctoral dissertation, Mälardalen University, Västerås, Sweden.

Saranya, C. M. and Nitha, K. P., 2015, 'Analysis of Security methods in Internet of Things', *International Journal on Recent and Innovation Trends in Computing and Communication*, vol .3, no. 4, pp. 1970–1974.

Sawyer, R., 2011, 'The Impact of new social media on intercultural adaptation', University of Rhode Island. Available at http://digitalcommons.uri.edu/cgi/v iewcontent.cgi?article=1230&context=srhonorsprog.

Thoke, V. D., 2015, 'Theory of distributed computing and parallel processing with applications, advantages and disadvantages', *International Journal of Innovation in Engineering, Research and Technology*, vol. 80, p. 9915. http://www.ijiert.org/ admin/papers/1452798652_ICITDCEME%E2.

Vongsingthong, S. and Smanchat, S., 2015, 'A review of data management in internet of things', *Asia-Pacific Journal of Science and Technology*, vol. 20, no. 2, pp. 215–240.

12

Analytical Approach to Sustainable Smart City Using IoT and Machine Learning

Syed Imtiyaz Hassan and Parul Agarwal

CONTENTS

12.1 Introduction to Smart City

The idea of the high-tech smart city is moderately new and it will soon take over from the knowledge city, the digitalised city and the feasible city. Although it has been used regularly, and exclusively after 2012, in recent years there has been a lack of unanimity in the conversation about its architecture and thought. The term "smart city" has covered various sub-themes such as smart urbanisation, smart financial systems, sustainability and the smart environment, high-tech technology, energy conservation, smart mobility, smart Medicare, and so on (Bolívar 2016). With the advance in Internet of Things devices and their application, the mobile spectrum, broadband communication, next generation and cloud-based technology, etc., informatisation inclines toward higher smarter phases. Recently some corporate houses have issued the concept of the "smart globe." The belief behind "smart globe" is that sensing devices should be embedded in public places like public transport stations, bridges, subways, roads, buildings, rainwater harvesting systems, dams, mercantile equipment and healthcare devices, and then physical facilities can be established, so that digital technology extends into the physical world, creating an "Internet of Things." Moreover, the Internet of Things may be connected with networking to enumerate the social community and physical systems. Human beings, machines, electronic devices can be managed in the incorporated system through networking and cloud computing, so that human invention and human life can be managed more accurately and energetically in smarter way, resource usage and productivity can be increased and the association between people and the environment improved. The smart (high-tech) city also relies on the IoT, sensing devices and cloud computing to provide an extensive network of linked electronic devices. Smart sensing devices and large-scale data analytics enable the move from the Internet of Things to real-time control. Internet of Things technology is applied in a smart city as a key component. For instance (see Figure 12.1), to offer user-friendly services, there is data capture, sharing and use by digital home appliances including refrigerators and washing machines in a smart home environment (Bunce 2016).

As is clear from its name, the Internet of Things implies connecting all sensing devices to the Internet. The Internet of Things is a peer-to-peer network that consumes standard networking protocols while its convergence point is the Internet. The key idea of the Internet of Things is the general presence of objects that can be measured, inferred, implied and that can change the environment. On this basis, the Internet of Things is enabled by the development of various objects as well as communication technologies. The Internet of things involved electronic devices including cell phones and other objects like food stuffs, home appliances, landmarks, monuments, works of art that can cooperate together to provide a common target (Castelnovo, Misuraca, and Savoldelli 2016). Mobile agents (software agent technology) have some advantages that are more useful for the Internet of Things. Mobile agents help to minimise the networking capacity, overcome network latency, and encapsulate protocols. In accumulation, they can execute autonomous tasks that would otherwise need broad configuration. On the down side, many agent systems require use of specific technologies and host environments, which complicates using them and implies that special skills are needed in creating agents. Mobile agents are executable (run-time) entities that can move from

IoT Application Domains

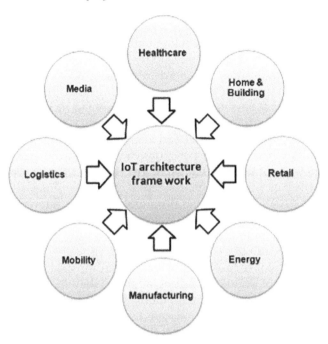

FIGURE 12.1
IoT Application domain

one host to another together with the internal state of the software application. It means that an executing agent may halt its execution on the current system and continue on another system. In fact, mobile agents represent a special case of moving code combining remote evaluation with preservation of the internal state (Bolívar 2016). The mobile agent technology is executed on a platform that is wholly distributed and fault tolerant to make the whole concept stronger. The characteristics of decentralisation and load distribution among different kinds of multi-source devices could interact with each other on a peer-to-peer system to provide services to clients. These functionalities may include day-to-day tasks such as controlling electronic appliances remotely and intelligently or navigating around a remote location.

12.2 Background

In the literature there are more than enough definitions of smart city, but the universal definition has not been framed yet. It seems that "digital" and "smart" city are both terms frequently used in educational research to delineate the intelligence of a city. How a city is said to be smart is covered in different research publications. What are the characteristics of a smart city? There are researchers from almost every field who mainly focus on the sustainable development of cities with a heightened reliance on a permanent progressive route map for their development. A setup to consider the model of smart cities is built on eight significant features: people and communities, leadership and organisation, advanced machinery, policy context, control, economy, infrastructure and natural surroundings (Anagnostopoulos et.al 2017) . These factors act as the foundation of a merged framework that puts forward guidelines and outlines to governments so as to be able to envisage smart city projects. The model and connections of the smart city with the digital city are based on the core content of the implementation procedures, the challenge of the ability to build and the impact of its progress. The basis of the smart city is the primary step to set up the smart city skeleton and the overall structural design; its progressive configuration and the areawise precision of tracing its properties as a foundation of the development of smart city planning, united with the entire amenities and organisation allied to the smart city agenda (Bandyopadhyay et.al 2011).

In Berrone et al (2018), the authors put forward an approach that can be used for the computation of smart city indices, although they chose indicators that are heterogeneous, and include vast amounts of information. The paper is concerned with the calculation of allocated weights and the approach used here is totally based on fuzzy logic for the considered indicators. The authors (Gandhinagar Municipal Corporation 2015) presented the paper with a new concept of ubiquitous-eco-city (u-eco-city) and concentrate on whether u-eco city is an astounding smart and sustainable city that represents an ultimate twenty-firstcentury city model or just a branding deception. Eco-city proposals what is aspired to in the twenty-first century. Eco-cities are considered as zero-carbon, carbon-neutral, low-carbon, ubiquitous-eco and sustainable, highlighting their position regarding sustainability.

This study focuses on the u-eco-city. In Grazia Concilio, and Marsh (2016), the authors discussed in detail the favourable circumstances by means of ICT and considered it to be a promising technology to lessen the use of energy in our cities. The authors developed an analytical model in which a study of ICT opportunities is pooled with a typology of domestic functions, i.e., all the daily activities that require energy. The energy that is required for domestic functions is evaluated with the help of a consumption-based point of view. De Jong et al. (2015) propose a theoretical agenda to study and examine two principal cases from the US and Asia. The main objective of this paper is to focus on the progress in building an efficient smart city by amalgamating a variety of practical viewpoints with a contemplation of smart city characteristics. Here a model is developed to carry out case studies probing how smart cities were being put into practice in San Francisco and Seoul Metropolitan City. The study's experimental outcomes imply that efficient, sustainable smart cities appear by virtue of methods that are influential in which communal and economic sectors are brought together in order to coordinate their activities and resources on an open advanced policy.

In Douglas (2016), the author developed a conceptual model that makes it simple to interpret how neighboring administrations build up demand-side guidelines tools that inspire the growth and dispersal of sustainably driven transformations that boost limited fiscal growth. The authors in Foster and Iaione (2015) give a broad understanding of the idea of the smart city while elaborating the categorisation of relevant application domains, such as: buildings, accommodation, government, natural assets and human dependence on these assets, transportation and mobility, the economy and people. They also investigated the dispersal of smart proposals by means of an experimental study intended to explore the extent to which an area is covered by a city's performance using the full potential of smart projects, accepting the function that various demographic, geographical, trade and industry and municipal variables may be affecting the move to build a smarter city. In this study, results reveal that the advancement patterns of a smart city are extremely dependent on its concerned circumstances. In Zaslavsky, Perera, and Georgakopoulos 2012), the authors studied previous and current programs for sustainability and the conditions suitable for living in cities and monitored indicators that are required for effective monitoring. In this study, the development of synthetic indices was proposed by means of principal component analysis and the real-time data is used as an alternative to historical data as the essential information that one may use to build a set of indicators to elucidate the proposal.

In an article, Gusmeroli and Haller (2009) summarise and give a detailed introduction regarding eco-town-based urban-development implantation in a city in north western Europe. The authors emphasised the development of eco-towns and the rules and regulations, processes and models which helped the eco-towns get started.

In Chang, Chiu, Ramachandran, and Li (2016), the main focus is to find the role of innovative, intelligent urban machinery and tools in the development of smart cities. The authors put forward a comprehensive review regarding the applications of a smart city framework while investigating the emerging practices of ubiquitous eco-cities. This paper (Internet and Things 2018) highlights the differences in public transport between Newcastle in the United Kingdom and the southern Brazil city of Florianopolis. The study made a comparison of models, and measured the progressive growth of Florianopolis as a sustainability model in South America. Hence, a detailed investigation of changes, models that were built and similarities and differences have been verified in order to discover the behaviors that direct social and political processes in the field of urban sustainability. This paper gives a proportional depiction of social and financially viable indicators like gross domestic product (GDP) per capita, price increases, employment, as well as the historical population evolution of the two cities.

In Qiu et al. (2018), the authors performed a relative case-study of three Asian cities, such as Penghu in Taiwan, Seoul in South Korea and Tianjin in China and revealed the special effects of multiple nationwide methods to eco-city progress. In their study, they compared the Asian cities with two European cities, Freiburg (Germany) and Sams (Denmark). The investigation makes out four transforming antecedents of the growth of an eco-city in Asia, which are (1) the presence of a dedicated local public authority, (2) deployment of a countrywide approach and policy, (3) a combination of national potential, and (4) commercial activity and the continuous engagement of local citizens. In Santoro et al. (2018), 16 sets of city measurement frameworks (8 smart-city and 8 municipal sustainability-evaluation models) were composed of 958 indicators in total by separating the indicators into 3 groups and 12 divisions. The paper put forward a comparison of modern technologies and smartness that must be implemented to make a city smart to urban sustainability frameworks. A common motive of smart cities is to improve sustainability with the help of technologies. In Qiu, Qiao, and Wu (2017), an indicator agenda for the estimation of a low carbon city (LCC) was created from the viewpoint of energy patterns, society and livelihoods, economics, inner-city mobility, solid waste, carbon and environment and water. A complete assessment scheme was engaged for LCC ranking by means of the entropy weighting factor method. The standard standards for LCC documentation were also known. The model was proposed to ten international cities in order to grade them according to their low carbon levels. A comparative study, made at multiple points of trade and industry, societal, and green development, improved the entire study. The outcome proved that Stockholm, Vancouver, and Sydney held a position higher than standard, signifying these cities gained a high grade value in the growth of low-carbon emissions.

12.3 Smart City Architecture

Defining smart city architecture is an arduous task, though researchers are working earnestly to set a standard for real-world formation. Apart from theoretically, it is difficult to define a common smart city architecture for real-world implementation, due to the alteration in appropriate features. Somehow researchers define a common architecture that consists of a majority of required features; according to Nathali, Khan, and Han (2018) smart city architecture consists of four layers i.e. sensing layer, transmission layer, data management layer and application layer as shown in Figure 12.2.

Security of the data is the main challenge of any smart city, which is why security modules have been integrated into each layer. The sensing layer collects the data from the physical layer which is the backbone of smart city

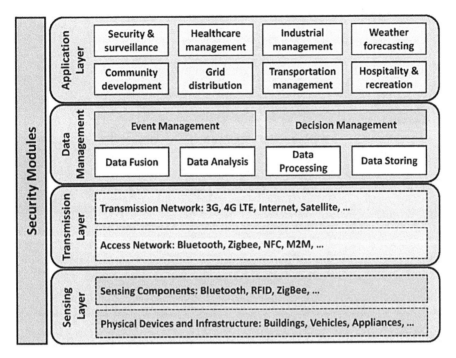

FIGURE 12.2
Smart city architecture. (Nathali, Khan, and Han 2018.)

architecture. The transmission layer advances the data to the upper layer. The data management layer processes and stores valuable information that is useful for the service provision offered by various applications at the top layer.

12.3.1 Sensing Layer

In reality, the smart city consists of several real-world elements, such as big data, complex computations, intelligent decision-making capabilities and more. According to Rong et al. (2014) data and computations play an important role in the implementation of the smart city as they help in decision-making.

Data collection is the most challenging as well as the most important task in the smart city environment; it is challenging because of heterogenous data and it is important because it controls rest of the operations of a smart city.

As shown in Figure 12.2, the sensing layer is a collection of smart data capturing devices that capture any data from any device (Deshpande et al. 2004). The sensing layer uses multiple hardware to sense data coming from different environments, such as RFID sensors, actuators, GPS, cameras, Bluetooth and Zigbee. These devices collect different data parameters, for example temperature, humidity, pressure light etc. (Bandyopadhyay and Sen 2011).

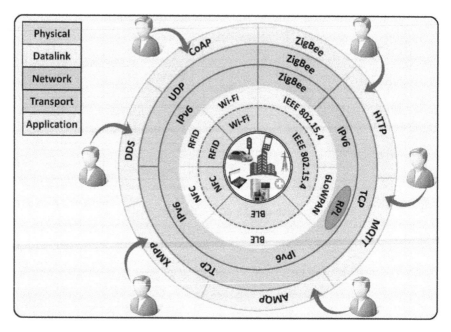

FIGURE 12.3
Communication technologies and protocols used in smart city deployment. (Nathali, Khan, and Han 2018.)

12.3.2 Transmission Layer

This layer acts as a bridge between the sensing layer and the data management layer, taking data from data sources and connecting with the management layer for further processing. The transmission layer uses various wired, wireless and satellite technologies. In terms of coverage, the transmission layer is divided into two parts, i.e., access transmission for short range, and network transmission for long range. NFC, Zigbee, M2M, Zwave and Bluetooth, are examples of access transmission, and 3G, 4G, 5G and low-power wide area networks (LP-WAN) are examples of network transmission. Figure 12.3 illustrates various communication technologies and protocols used in smart city deployment.

12.3.3 Data Management Layer

The second-last layer of probable smart city architecture is the data management layer, also known as the brain of the smart city. The main moto of this layer is to perform a variety of operations on data, like data manipulating, organising, analyzing, storing and decision-making. In fact, for sustainable smart city efficiency of the data, the management layer is crucial; also the essential duty of the data management layer is to manage, review, clean and maintain the data, since service performance of smart city operations rely on data management.

12.3.4 Application Layer

This is the only layer that directly interacts with urban citizens, being the topmost layer of smart city architecture. This layer should be highly intelligent so as to offer intelligent services to urban citizens, because smart city citizens are very concerned about the intelligent behavior of the application layer, for example in weather forecasting.

The application layer offers a variety of services including smart transportation, weather forecasting, community development, grid distribution, etc. The application layer boosts smart city performance with a number of applications that work in combination, because developing isolated smart application has less impact over those applications that utilise processed and stored data, work with majority of applications enhances the performance of smart city operations, the key point being that those applications share the information among different applications, which seems a promising approach for the evaluation of smart cities.

12.4 Major Smartest Cities in the World

With smart hospital services to a greener atmosphere, from smart transportation to a smart grid, smart cities are quickly revolutionising the way we think about urban living. In this section we list the top ten smart cities in the world as per IESCE Cities in Motion Index 2018 survey (Berrone and Ricart 2018). You may also refer to Figure 12.4 which ranks the top 50 smart cities in the world.

12.4.1 Reykjavik

The city of Reykjavik is the capital of Iceland. The city has been particularly recognised for its environmental initiatives. For example, the city encourages citizens to use public transport more frequently, for which purpose government recently developed a public transportation app for city buses in the greater Reykjavík area called Straetó. The Icelandic capital has tried to involve the public in its plans through Better Reykjavik, an online consultation forum where citizens can present their ideas on the services and operations of the city.

12.4.2 Tokyo

Tokyo is the number one smart city in the Asia-Pacific region, and gained fourth position in the list of best smart cities in the world, according to the IESE index. As one of the most popular metropolitan areas in the world with a high rate of labour productivity, human capital and economy are specialities

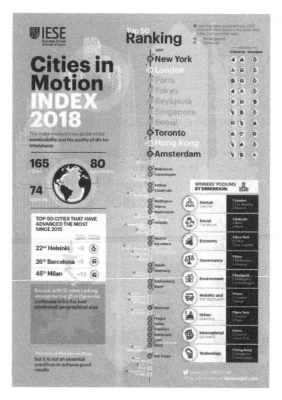

FIGURE 12.4
Top 50 smart cities in the world. (From Berrone and Ricart 2018.)

of Tokyo. Tokyo is all set to host the Olympic Games in 2020. Tokyo will use face recognition technology to improve security whilst driverless taxis are expected to ferry athletes and tourists from place to place.

12.4.3 Paris

Paris, the French capital, has been ranked third. The city has been appreciated for its efforts in international outreach, transportation and mobility. The city is currently in the process of developing 127 miles of metro line consisting of 68 new stations. It is projected that by 2050, the city will also replace the entire 4,500-bus fleet of the RATP (the Paris Region's primary public transport operator) with electric or natural gas vehicles (NGVs).

12.4.4 London

IESE ranked London as second-best smart city in the world. As per the report it is considered to be one of the best capitals. It was also recognised for its mobility and transportation, international outreach, economy, governance, technology and urban planning.

12.4.5 New York City

New York City is considered to be the most developed, equipped and advanced smart city in the world. It has great focus and conservation of resources. The city has automated various services while keeping sustainability in mind. Almost all the major services such as smart dustbin, smart energy monitoring, water billing and conservation using the latest technologies.

12.5 Role of Fog Computing in Smart City

Smart city is all about IoT or connected sensors promising to deliver extremely fast services to citizens. These sensors generate numerous amounts of data that are mostly latency sensitive. Fog computing plays an important role to ensure timely and stable data flow among millions of connected devices in the smart city, particularly for applications performing critical functions and demanding quick responses.

Cisco integrated fog computing modules into smart cities (Moffett, Marcus, and Dodson 2019), for example:

- Fog computing improved operations and services for water, sewer, energy, waste and fleet management.
- Fog computing enhanced public safety for EMS/fire/police dispatch and natural/manmade disaster response.
- Fog computing improved transportation using direct vehicle-to-infrastructure (V2i) and vehicle-to-vehicle (V2x) communications/actions to change signage, alter traffic signals, optimise flow and control autonomous vehicles.
- Embedded connected devices administer and ensure timely, correct treatments in healthcare and social services.
- Strengthened law enforcement is implemented through augmented reality (AR) glasses that quickly identify known or wanted criminals through facial recognition.

Fog computing modules can benefit in terms of handling issues that citizens face in everyday life, such as waste, parking, public safety lightning, etc.

12.6 Analytical Approach to Sustainability in the Smart City

The United Nations Sustainable Development Goals (SGDs), 2030, has emphasised and defined sustainability for future generations ("United Nation Development Programme: Sustainable Development Goal" 2019). That is why one of the major objectives of the smart city will be sustainability and environmental protection along with other goals such as increased resource efficiency and ease of use. The analytical approach to sustainability in the smart city explores the applications, propositions, and implementations of advanced ICT that makes the city intelligent with special focus on sustainability. The applicability of recent available enabling technologies that may be helpful in achieving sustainability must be analyzed. Some of these available technologies are machine learning, IoT, edge computing, cloud computing, big data and others. Though the applications of these technologies in smart city services is challenging, they may contribute to a significant improvement in sustainability if applied intelligently. Realising the growth of smart cities in the world, with the multidimensional applicability of enabled technologies, the UN has studied and published "Big Data and the 2030 Agenda for Sustainable Development" ("Big Data for Sustainable Developmen," n.d.).

12.7 Enabling Technologies for Sustainability

There are various forces that lead to the emergence of a concept and then guide it to growth. One of the leading factors is the market. The smart city is no exception. While developing smart cities one has to consider resource conservation as the prime factor. Other factors such as economic development will be considered next.

12.7.1 IoT

The IoT is a communications archetype that envisages that everyday objects will be fitted with microcontrollers, digital communication transceivers and appropriate protocol stacks that enable them to interact with each other and with customers and become an essential component of the Internet (Muñoz et al. 2018). It has various applications in smart cities. Apart from various traditional applications in smart cities it can play a vital role in sustainability. Some of these are waste management (Anagnostopoulos et al. 2017), noise monitoring (Noriega-Linares et al. 2017), traffic congestion detection (Djahel et al. 2015), smart parking (Hu and Ni 2018), smart lighting (Daely, Reda, Satrya, Kim, and Shin 2017) and so forth.

Application of smart ICT in waste management can lead to substantial savings and economic and ecological benefits. For example, smart waste containers can sense load levels and optimise the collector truck route resulting in decrease in waste collection costs. The intelligent use of sensors as an IoT device may be used for noise monitoring for reducing acoustic pollution. The use of GPS, camera-based traffic surveillance systems, acoustic sensors and other IoT devices are very helpful in reducing traffic congestion by smart routing. It will be helpful in achieving sustainability by reducing the carbon footprint and air pollution. Smart parking is another sustainability solution of the smart city, with various sustainability advantages. It reduces the parking search time, and lowers carbon emissions and traffic congestion. In addition, the use of Radio Frequency Identifiers (RFIDs) or Near Field Communications (NFCs) helps in reserving slots for residents or disabled persons. Optimising street lighting systems through the use of IoT in smart cities helps reduce electricity consumption. The optimal intensity of street lamps based on the levels of natural light, climate, and the presence of people is another service that may be used.

12.7.2 Machine Learning

Machine learning plays an important role in imparting a sense of intelligence and sustainability in the smart city. Machine learning algorithms equip computers with the ability to learn without being explicitly programmed. There are different machine learning algorithms that may be used. Machine learning is used for managing uncertainty. It is used for building a model for data analytics. It may be used in the smart city as a tool for descriptive, predictive, diagnostic and prescriptive analytics. Smartness, in different sustainable cases such as waste management, noise monitoring, traffic congestion detection, smart parking and smart lighting, is possible with the help of machine learning algorithms.

12.7.3 Big Data

We are in the era of a data revolution. The mechanism of data collection by crowdsourcing, the available open data and the data collected from IoT devices are already transforming societies in terms of an abundance of data. The data as well as the rate of production of data is so huge that it cannot be managed by traditional approaches and hence is termed big data. There is no fixed definition of big data. Many definitions have been suggested; all have one thing in common, that is that data is large, unstructured and the production rate is very high. The smart city uses this big data for intelligence and decision-making. It is necessary for these cities to manage data in a planned manner in order to contribute to sustainable development goals.

12.8 Proposed Model for the Analytical Framework of a Sustainable Smart City

An analytical framework as depicted in Figure 12.5 uses the previously mentioned key enabling technologies as well as some other technologies for

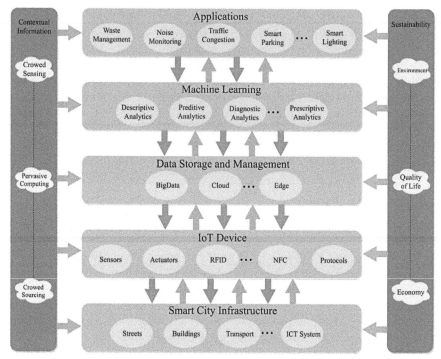

An analytical framework for sustainable smart city

FIGURE 12.5
An analytical framework for the sustainable smart city.

smart city sustainability. These are, but not limited to, cloud computing, edge computing, fog computing, pervasive computing and crowd sensing. Cloud computing supports the idea of rapid production, storage and consumption of information. Open data on a cloud is used for sharing inter-functional data. Fog and edge computing with analytics are used for pre-processing the data at source and are then distributed to the cloud or services to be consumed, which helps achieve sustainability by saving data traffic. It also helps optimise the latency of decision-making process for connected devices. Pervasive computing and crowd sensing utilise the smart devices of the general population, such as mobile phones, as computing resources.

12.9 Conclusion

A lot of research has been done dealing mainly with smart cities providing citizen services for an improved quality of life using data analytics and business intelligence. The domain of sustainable smart cities is still in the early stages of development but has gained some footholds. This research is an attempt to provide an analytical approach to improve sustainability in smart cities using the IoT and machine learning. To achieve sustainability, smart cities rely heavily on technologies and novel algorithms that deal with the intelligence of smart objects leading to efficient decision-making. Toward these ends, the IoT and machine learning may play a vital role for energy efficiency leading to the mitigation of negative effects on the environment. The types and management of sensors, actuators, communication protocols, numbers of layers and algorithms used by the IoT to connect tools, people, places and machines are considerable factors for sustainability. Since a smart city has a huge collection of smart objects, it has to deal with massive data for obtaining meaningful information. Hence, the machine learning algorithms are important for sustainability. Various applications such as waste management, noise monitoring, traffic congestion detection, smart parking, smart lighting and others are a few such systems that are helpful in achieving sustainability in the smart city. Key enabling technologies for sustainability – IoT, machine learning, big data, cloud computing, fog and edge computing, crowd sensing, and pervasive computing – have been discussed. Finally, this chapter concluded with an analytical framework for sustainable smart cities using the IoT and machine learning.

References

Anagnostopoulos, T., Zaslavsky, A., Kolomvatsos, K., Medvedev, A., Amirian, P., Morley, J. and Hadjieftymiades, S., 2017, 'Challenges and opportunities of waste management in IoT-enabled smart cities: a survey', *IEEE Transactions on Sustainable Computing*, vol. 2, no. 3, pp. 275–289. doi:10.1109/tsusc.2017.2691049.

Bandyopadhyay, D. and Sen, J., 2011, 'Internet of things : applications and challenges in technology and standardization', *Springer Science+Business Media, LLC.*, vol. 2011, pp. 49–69. doi:10.1007/s11277-011-0288-5.

Berrone, P. and Ricart, J. E., 2018, 'IESE cities in motion index 2018', *IESE Insight*. doi:10.15581/018.ST-471.

"Big Data for Sustainable Development." n.d., United Nations. Available at https://www.un.org/en/sections/issues-depth/big-data-sustainable-development/index.html.

Bolívar, M. P. R., 2016, 'Mapping dimensions of governance in smart cities', in *17th international digital government research conference on digital government research*, ACM, New York, pp. 312–324. doi:10.1145/2912160.2912176.

Bunce, S., 2016, 'Pursuing urban commons: politics and alliances in community land trust activism in East London', *Antipode*, vol. 48, no. 1, pp. 134–150. doi:10.1111/anti.12168.

Castelnovo, W., Misuraca, G., and Savoldelli, A., 2016, 'Smart cities governance: the need for a holistic approach to assessing urban participatory policy making', *Social Science Computer Review*, vol. 34, no. 6, pp. 724–739. doi:10.1177/0894439315611103.

Chang, V., Chiu, D. K., Ramachandran, M., and Li, C. S., 2016, 'Internet of things, big data and complex information systems: challenges, solutions and outputs', in *COMPLEXIS 2016 and CLOSER 2016 selected papers and CLOSER 2015 Keynote,* IoTBD.

Concilio, G., Marsh, J., Molinari, F. and Rizzo, F., 2016, 'Human smart cities: a new vision for redesigning urban community and citizen's life', Springer, Cham, vol. 364 (February 2016), pp. 269–278. doi:10.1007/978-3-319-19090-7.

Daely, P. T., Reda, H. T., Satrya, G. B., Kim, J. W. and Shin, S. Y., 2017, 'Design of smart LED streetlight system for smart city with web-based management system', *IEEE Sensors Journal,* vol. 17, no. 18, pp. 6100–6110. doi:10.1109/JSEN.2017.2734101.

Deshpande, A., Guestrin, C., Madden, S., Hellerstein, J. and Hong, W., 2004, 'Model-driven data acquisition in sensor networks', in *Proceedings 2004 VLDB Conference,* pp. 588–599. doi:10.1016/b978-012088469-8/50053-x.

Djahel, S., Doolan, R., Muntean, G. M., and Murphy, J., 2015, 'A communications-oriented perspective on traffic management systems for smart cities: challenges and innovative approaches', *IEEE Communications Surveys and Tutorials,* vol. 17, no. 1, pp. 125–151. doi:10.1109/COMST.2014.2339817.

Douglas, G. C. C., 2016, 'The formalities of informal improvement: technical and scholarly knowledge at work in do-it-yourself urban design', *Journal of Urbanism,* vol. 9, no. 2, pp. 117–134. doi:10.1080/17549175.2015.1029508.

Foster, S. and Iaione, C., 2015, 'The city as a commons', *SSRN Electronic Journal,* pp. 281–349. doi:10.2139/ssrn.2653084.

Gandhinagar Municipal Corporation, 2015, *Presentation on Gandhinagar* - Smart City, edited by Renata Paola Dameri and Camille Rosenthal-Sabroux, 1st ed., Springer International Publishing, Switzerland. doi:10.1007/978-3-319-06160-3.

Gusmeroli, S., Haller, S. and Harrison, M., 2009, 'Vision and Challenges for Realizing the Internet of Things', in H. Sundmaeker, P. Guillemin, P. Friess and S. Woelfflé (eds.), *Proceedings of the 3rd STI roadmapping workshop,* Vol. 1, Brussels. doi:10.2759/26127.

Hu, L. and Ni, Q., 2018, 'IoT-driven automated object detection algorithm for urban surveillance systems in smart cities', *IEEE Internet of Things Journal,* vol. 5, no. 2, pp. 747–754. doi:10.1109/JIOT.2017.2705560.

Hu, J, Yang, K., Marin, S. T., et al., 2018, 'Guest editorial special issue on internet-of-things for smart cities', *IEEE Internet of Things Journal,* vol. 5, no. 2, pp. 468–472. doi:10.1109/JIOT.2018.2792885.

Jong, M. D., Joss, S., Schraven, D., Zhan, C. and Weijnen, M., 2015, 'Sustainable-smart-resilient-low carbon-eco-knowledge cities; making sense of a multitude of concepts promoting sustainable urbanization', *Journal of Cleaner Production,* vol. 109, pp. 25–38. doi:10.1016/j.jclepro.2015.02.004.

Moffett, M. and Dodson, K., 2019, 'Understanding the value of edge and fog computing to your community', *Cisco Blogs.* Accessed September 15. Available at https ://www.cisco.com/c/en/us/solutions/industries/government/us-government -solutions-services/fog-computing.html.

Muñoz, R., Vilalta, R., Yoshikane, N., Casellas, R., Martínez, R., Tsuritani, T. and Morita, I., 2018, 'Integration of IoT, transport SDN, and edge/cloud computing for dynamic distribution of IoT analytics and efficient use of network resources', *Journal of Lightwave Technology,* vol. 36, no. 7, pp. 1420–28. doi:10.1109/ JLT.2018.2800660.

Nathali, B., Khan, M. and Han, K., 2018, 'Towards sustainable smart cities : a review of trends, architectures, components, and open challenges in smart cities', *Sustainable Cities and Society*, vol. 38 (January), pp. 697–713. doi:10.1016/j.scs.2018.01.053.

Noriega-Linares, J. E., Rodriguez-Mayol, A., Cobos, M., Segura-García, J., Felici-Castell, S. and Navarro, J. M., 2017, 'A wireless acoustic array system for binaural loudness evaluation in cities', *IEEE Sensors Journal*, vol. 17, no. 21, pp. 7043–7052. doi:10.1109/JSEN.2017.2751665.

Qiu, T., Chen, N., Li, K., Atiquzzaman, M. and Zhao, W., 2018, 'How can heterogeneous internet of things build our future: a survey', *IEEE Communications Surveys and Tutorials*, vol. 20, no. 3, pp. 2011–2027. doi:10.1109/COMST.2018.2803740.

Qiu, T., Qiao, R. and Wu, D. O., 2017, 'EABS: an event-aware backpressure scheduling scheme for emergency internet of things', *IEEE Transactions on Mobile Computing*, vol. 17, no. 1, pp. 72–84. doi:10.1109/tmc.2017.2702670.

Rong, W., Xiong, Z., Cooper, D., Li, C. and Sheng, H., 2014, 'Smart city architecture: a technology guide for implementation and design challenges', *China Communications*, vol. 11, no. 3, pp. 56–69. doi:10.1109/CC.2014.6825259.

Santoro, G., Vrontis, D., Thrassou, A. and Dezi, L., 2018, 'The internet of things: building a knowledge management system for open innovation and knowledge management capacity', *Technological Forecasting and Social Change*, vol. 136, pp. 347–354. doi:10.1016/j.techfore.2017.02.034.

"United Nation Development Programme: sustainable development goal," 2019, United Nations Development Programme. Accessed September 14, https://www.undp.org/content/undp/en/home/sustainable-development-goals.html.

Zaslavsky, A., Perera, C. and Georgakopoulos, D., 2012, 'Sensing as a service and big data', in International conference on advances in cloud computing, Bangalore, India. http://arxiv.org/abs/1301.0159.

13

Traffic Flow Prediction with Convolutional Neural Network Accelerated by Spark Distributed Cluster

Yihang Tang, Melody Moh and Teng-Sheng Moh

CONTENTS

13.1 Introduction

The prediction of traffic flow has been broadly applied in modern societies. Many industries use real-time data to generate instant predictions. Traffic flow prediction is important in quite a few areas, such as urban city design, tourism development, major event planning and personal travel and commuting arrangements. It is, however, a challenge as well as a painful process to find a balance between prediction accuracy and the cost of computation and time. Accurate results often need complicated analysis based on a large amount of data. Alternatively, one can sacrifice accuracy to save some cost, but it may lead to unacceptable prediction results.

For traffic flow forecasting, this chapter proposes a method that runs a prediction on images containing moving vehicles. By analyzing the movement of vehicles and the relationship between vehicles and locations, a classification of traffic situation among all available coordinates in the route is conducted, and it can display the overall traffic flow from the desired starting point to the destination.

This chapter further studies the characteristics and the performance of Convolutional Neural Networks (CNN) to explore how it may utilise a distributed environment. Specifically, three components of CNN are discussed for object recognition in the image: learning rate, activation function, and pooling layer. By adjusting these three components, the proposed CNN model becomes an effective tool to recognise vehicles from captured images, and to transfer them to classifications between the number of vehicles and coordinates.

The optimisation of CNN mainly aims to increase prediction accuracy. However, training of the model takes an enormous amount of time, because CNN are complicated neural networks (NN), and the size of the dataset is huge. To accelerate the entire training process, Apache Spark has been utilised. It is a cluster-computing framework that supports the MapReduce paradigm without depending on it (Manzi and Tompkins, 2016). Different modes of the Spark cluster are implemented and compared. The performance of GPU and CPU accelerations is also explored based on a series of experiments.

This chapter is organised as follows. Section 2 presents background and related studies. Section 3 introduces existing machine learning and deep learning models, as well as illustrates some preliminary results by comparing the performance of CNN with that of traditional machine learning methods. Section 4 describes the proposed solution: Distributed CNN running on Apache Spark cluster. Section 5 illustrates the performance evaluation. Finally, Section 6 concludes the chapter, and suggests some future research directions.

13.2 Background and Related Studies

13.2.1 Machine Learning

Walter Pitts and Warren McCulloch were the first researchers who tried to mimic the way a neuron was thought to work in the 1940s (Beam, 2017). This was considered the starting point of machine learning since it established a concept about how a machine could simulate the mechanism of a human brain. Then, ML (machine learning) and AI (artificial intelligence) entered its first winter. Not many breakthroughs happened until the 1970s, when Rosenblatt's perceptron began to gain a bit of attention (Beam, 2017). However, their study results were doubted by Seymour Papert, who proved that perceptron was incapable of learning the simple exclusive-or function. This led to the first AI winter.

Then, after over three decades of silence, Hinton in 2006 declared that he had found out how the brain actually worked (Beam, 2017). He introduced the idea of unsupervised pretraining and deep belief nets (Beam, 2017), which has brought back interest into this field. Soon, more and more papers about NN appeared and attracted more attention. In 2012, a significant breakthrough occurred. J. S. Denker (Beam, 2017), led a project that successfully used deep nets for speech recognition. Then, the use of graphics processing units (GPU) to train the model broke through the limitation of physical resources, and research on AI and ML literally ushered in a new era.

13.2.2 Deep Learning

Before DL (deep learning) came into view, ML struggled with handmade features. Obviously, this is not "intelligent" enough, but the appearance of deep learning solved this question. In general, the most important difference between traditional ML and DL is how features are extracted (Alom et al., 2018). Unlike traditional ML, DL learned and extracted features "automatically" and represented hierarchically in multiple levels (Alom et al., 2018). This means that DL is stronger than traditional ML in most practical scenarios.

13.2.3 CNN

CNN, as one of the most famous DL classes, is no doubt the superstar in visual imagery analysis. To pursue the origin of CNN, we need to go back to the 1980s, when Kunihiko Fukushima introduced the concept known as "neocognitron" (Fukushima, 2007). He introduced two significant components in CNN: convolutional layers and downsampling layers. These two components are still the core of CNN even after several decades.

13.2.4 Spark Cluster and Distributed Environment Acceleration

The development of Apache Spark is a journey from academia to industry. In 2009, a class project was proposed in University of California, Berkeley, which was about building a cluster management framework that could support different types of cluster computing systems (Madhukar, 2015). This project was called Mesos, and it is the basic element of Apache Spark.

Scalable and efficient LPR (SELPR) is an approach that uses YOLO network to detect the location of a vehicle license plate, which was introduced by Zhang et al. (2018). This approach is deployed on Apache Spark and takes advantage of it to improve the processing speed.

Zhang and other authors also used CNN to construct their model since it relates to object recognition in the image, and CNN is no doubt the best choice for this task. Similar to the project proposed in this paper, Zhang also used the Spark cluster to establish a distributed environment, but it was mainly used for data preprocessing. Although it indeed improves the performance of CNN model, we believe there is more potential to exploit.

13.3 Existing Machine Learning and Deep Learning Methods

Before we chose CNN, we did a lot of research on the traditional ML classifiers and DL model CNN. My research mainly focused on two aspects: the accuracy and time cost for a model/classifier to identify the specific object from images. The traditional ML classifier that we chose includes Decision Tree, Random Forest, SVM (Support Vector Machine) and KNN (K Nearest Neighbors).

13.3.1 Decision Tree

Decision Tree is a tree-like model, which is frequently used in the decision process. In general, this classifier mainly uses a conditional control statement to train data and extract features. Each node in the decision tree is a "test," and each test generates separate class labels. This classifier is considered expensive to construct and is fast. In the dataset, there exist a large number of records that have redundant or repeated attributes, and the decision tree is a good tool to handle this type of data.

13.3.2 Random Forest

Random Forest can be considered an advanced version of the decision tree, so sometimes people also called it a random decision tree. Compared to the decision tree, random forest constructs multiple decision trees randomly and takes the mean or mode to classify data while training data.

In this project, the random forest is also chosen because it usually provides a more accurate prediction result than the decision tree does, even though it also takes more time and computational cost. Besides that, the random forest provides a congestion state that cannot be achieved by using decision tree because it can reflect the traffic in an area rather than just a data point, which is very effective for this project (Liu and Wu, 2017).

13.3.3 SVM

Support Vector Machine (SVM) is a discriminative classifier that defines a hyper plane to categorise data. A hyper plane can be linear or non-linear, and since most roads in are not distributed linearly, the non-linear hyper plane is a better choice for this project. Another key concept is the kernel. A kernel is a function that implies the correlation of coordinates without computing it. This function can help with avoiding overfitting (Duan, 2018). In this chapter, we use a kernel function called Radial basis function (RBF). RBF takes two samples, x and x', as feature vectors in some input space, and defines the squared Euclidean distance based on it.

13.3.4 KNN

K nearest neighbor is another ML classifier that has been widely applied. K is the number of constant samples that the user defines. This method aims to "remember" the relative position of previous samples in multi-dimension coordination. A sample will be assigned to a class most common among its k nearest neighbors. KNN is relatively expensive compared to the decision tree or random forest. Further performance comparisons will be given in the following section. Another concept that affects how KNN categorises data is the way of "voting." The most common way to vote to decide which class a data point belongs to is the majority vote, which means that this data point will be categorised to the class that is major among k nearest neighbors. However, this is sometimes not the best fit, and a custom vote is an alternative method in that case.

A custom vote using Euclidean distance performs more accurately than the default majority vote in this dataset. One of the small green areas in the blue class is excluded by using the customised voting method.

13.3.5 CNN

CNN, as mentioned above, is a popular class of DL. CNN requires a conversion of the input: for image processing, the raw image needs to be converted to a matrix of pixels. Based on image resolution, the height, the width and the dimension of the matrix are various. Then, these matrices of pixels are passed into a structure called the hidden layer, which typically contains convolutional layers with activation functions as filters, a pooling layer, which is also known as a subsampling layer and a fully connected layer, which sometimes come with another pooling layer. Then, the final result is passed to the output layer. In this project, the design and optimisation of CNN model also follow this structure. Details about the modification on the model will be provided in the following section.

13.3.6 Comparison of Performance

Now, the question is: Which of the above tools has the better performance in the case of detecting an object in the image? The best way to answer this question is to set up an experiment and compare them directly. Therefore, we implemented all ML classifiers using a python library called scikit-learn and implemented CNN using a framework called deeplearning4j. Figure 13.1 and Figure 13.2 show the test result based on a sample containing 5000 frames in 3, 6, 9 and 12 months.

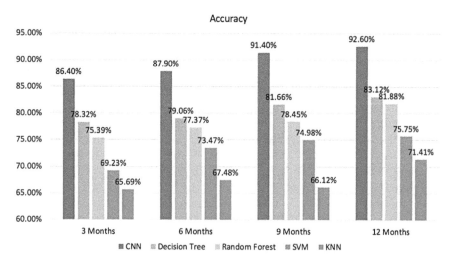

FIGURE 13.1
Accuracy –- 5 ML classifiers and CNN.

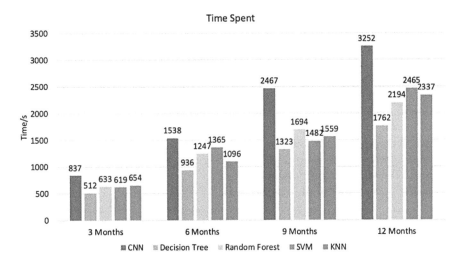

FIGURE 13.2
Training Time – 5 ML classifiers and CNN.

It is obvious that CNN generates a much more accurate prediction than all five ML classifiers. This is because it has a more complicated structure and avoids the bias caused by predefined classes. Unfortunately, CNN takes much more time to train and to get the final results. As shown in Figure. 13.2, the time CNN spends on a sample containing frames in 12 months is almost twice as much as Decision Tree does. This leads to the core of this project: how to accelerate the training process of CNN. As stated earlier, the answer is Spark cluster, and a detailed explanation about how to do it is given in the following section.

13.4 Proposed Method: CNN with Spark

13.4.1 Workflow

After evaluating the performance of CNN and the traditional ML methods, it is not hard to understand that CNN has excellent accuracy on object detection in the image. However, how to convert this advantage of CNN into the prediction of traffic flow is also a challenge. In the CityCam dataset, the location of each camera is given. Then, if we connect the detected number of vehicles with the location of each camera, we can easily get an estimation of the number of vehicles between two points on a map. Besides that, when we consider this question, the number of vehicles is not the only factor that affects traffic. The situation of the road also does. Thus, instead of simply detecting how many vehicles are in an image, we tried to get a percentage that represents how many pixels belong to detected vehicles in the whole matrix that converted from an image. Then, no matter the width of a road, this percentage can still display the traffic situation at that time. Thus, the workflow of making a prediction is: (1) CNN obtain the percentage that represents the region belonging to a detected vehicles/the size of the whole image, (2) each point that represents a camera from the starting point to the destination is assigned with a percentage that is the average of the percentage obtained in Step (1), (3) the final prediction between two points on a map is formed by each camera with a percentage on the route. Details about how to get the percentage of an image are provided in the next section since it relates to the specific component of CNN.

Thus, the workflow of making a prediction is as follows:

1. CNN obtains the percentage that represents the region belonging to detected vehicles/the size of the whole image.

2. Each point that represents a camera from the starting point to the destination is assigned a percentage that is the average of the percentage obtained in Step (1).

3. The final prediction between two points on a map is formed by each camera with a percentage on the route. Details about how to get the percentage of an image are provided in the next section since it relates to the specific component of CNN.

13.4.2 CNN Model Design and Modification

13.4.2.1 Learning Rate

Learning rate is an important hyper-parameter that affects the adjusting of the weight of the NN with respect to the loss gradient (Zulkifli, 2018). In other words, the learning rate is the most important factor that decides the rate of weight changing in the NN model. A naïve NN usually takes a stable learning rate when it is configured. However, if you use one constant learning rate from beginning to end, one of the shortcomings is that the gradient loss can be inefficient.

The relationship between gradient loss and the learning rate changes along with the increase of epochs. So, a constant learning rate cannot guarantee the best results all the time.

We performed an experiment to explore the relationship between learning rate and gradient loss for this project, and Figure 13.3 shows the result.

Figure 13.3 shows the loss of gradient decent of three different learning rates: 1.00E-01, 1.00E-03 and 1.00E-06. Obviously, there is no learning rate that is the best choice all the time, so a better strategy is to adjust the learning rate to achieve the best combination. In this project, starting with learning rate equals to 1.00E-01, then switching to 1.00E-03 after the training of the third epochs has finished, and changing to 1.00E-06 after the training of

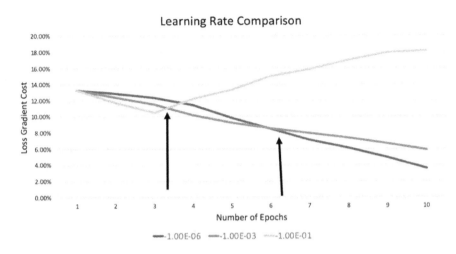

FIGURE 13.3
Comparison of loss gradient decent between different learning rates.

the sixth epochs has finished, is the best strategy. The rate of gradient loss is kept at the lowest rate even though the learning rate changes from fastest to slowest.

13.4.2.2 Activation Function

Activation function is the function that converts the linear output generated from convolutional layer to a non-linear result. For example, the convolutional layer generates a series of values from 0 to 10, an the activation function basically says: "everyone above 5 belongs to class 'yes' and the rest belongs to the class 'no'." Then, all values are classified into two complimentary classes.

In almost all of CNN, ReLU (Rectified Linear Unit) is used as the activation function. Of course, compared to other activation functions, such as Sigmod, ReLU is popular for a reason: ReLU can reduce the likelihood of vanishing gradient. Before we further explain these two benefits, recall that the definition of a ReLU is $h=\max(0, a)$ where $a=Wx+b$. For the reason mentioned earlier, when $a>0$, the gradient has a constant value. On the contrary, the gradient of Sigmoid becomes increasingly small as the absolute value of x increases. Therefore, the constant gradient of ReLU results in faster learning. Meanwhile, this characteristic also leads to less run time, which no doubt can speed up the training process. Although ReLU can potentially cause "deader" cells due to its nature when $a<0$, it is still the activation that is closest to "perfect" thus far.

13.4.2.3 Pooling Layer

Similar to the convolutional layer, the pooling layer is also responsible for reducing the spatial size of the features extracted from the input matrix. The benefit that the pooling layer brings is to reduce the cost of the resource. More specifically, the most straightforward purpose of the pooling layer is to extract domain features within part of the input matrix. There are many ways to achieve this. The most popular method is called Max Pooling. Basically, Max Pooling takes the maximum value from each partition of the general input and reconstructs a matrix that contains only the maximum value extracted.

In a model with stride equal to 2 and 2*2 filters, the result may vary based on different combination of stride and the dimensions of the filter. In this project, max pooling is also the first choice that we made, and it performed well. However, its performance is not as "perfect" as we expected. Therefore, we constantly changed the sample data and ended up discovering that the problem is the overlapping between vehicles. When two vehicles overlapped in an image, the selection of major features ignores part or even all of one vehicle. This leads to high standard deviation depending on the partition of the dataset.

In order to solve this problem, another type of pooling layer, SoftMax was introduced. SoftMax still classifies features based on the input, except that it assigns each feature a confidence level rather than taking only the maximum value. This means that SoftMax, compared to Max, changes from the top feature extracted from a specific region to the most possible top feature extracted from a specific region. A threshold is set, and features that have confidence levels higher than the threshold are counted; the possibilities of all qualified features are added. All features that have a confidence level larger than the threshold can be considered as "liked" and the rest are features that the model does not like in this image.

Image pixels are stretched into a column and a matrix multiplication is performed to get the score for each class. Then the weight of each weight is adjusted based on the score obtained.

However, nothing is perfect. SoftMax uses more resources since it may keep track of multiple classes instead of only one maximum value Thus, using SoftMax all the time is not advised. My method is to use Max after the first convolutional layer and activation function and to use SoftMax after the fully connected layer. This strategy, in our opinion, combines the strength of two pooling functions and sets off their disadvantages.

13.4.2.4 Final CNN Model

Figure 13.4 is our final CNN model. As shown in the figure, this is a 7-layer CNN model. The first four layers are the convolutional layer with the RELU and pooling layer using Max and SoftMax. Then, a flattened layer is implemented to convert extracted features to classes. Last, but not least, a fully connected layer is implemented to group pixels into different classes, and an additional pooling layer is implemented to finish the final filtering.

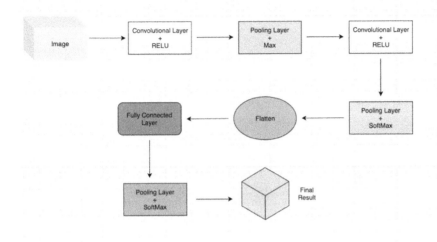

FIGURE 13.4
Final CNN model.

13.4.3 Spark Cluster Configuration

After optimising the model itself, a more important and practical question is that of speed. As introduced earlier, the DL model is far more complicated than traditional ML classifiers, so it takes much more time to obtain results. In fact, before we implemented any optimisation, the training of the model in this project took over ten hours to get a result with accuracy of more than 90%, which is not acceptable. Thus, it is necessary to find a way of accelerating the training process. Spark cluster is the answer we found. It provides a distributed environment that allows the model to be trained in parallel. It is a very good solution that efficiently allocates hardware resources to improve the performance of software.

One important challenge of deploying CNN in a Spark cluster is that it converts the entire training project into a series of independent sets of processes and runs all of the jobs on each worker node concurrently. Therefore, how to manage these nodes so that they can work more efficiently on CNN training tasks is the core problem when we configure the spark cluster.

13.4.3.1 Four Modes

So far, the definition of the distributed cluster is still ambiguous. In other words, there are many ways to construct a distributed Spark cluster and many of them have been proved to be good choices in specific fields. In other to find and compare the best mode for this project, the four most popular modes of Spark are implemented:

- Single server – CPU
- Single server – GPU
- Single server – Multi-GPUs
- Multi server – Single GPU per Server

13.4.3.2 Memory Layout

In Spark, each worker node has physical resources to use, like CPU, GPU, Memory, etc. Memory is one of the crucial components that directly relates to the performance of each worker node. Due to the consideration of fault tolerance, partial memory is restricted to use while an application is running on the node because they are preserved for failure recovery. In Spark, the size of this type of memory can be adjusted, which means we can choose how to partition the memory and find a balance between performance and fault tolerance. DeepLearning4j sets a boundary for the adjustable memory, which is 3.5 GB / 4 GB, so we choose types of memory layout – 3 GB / 4GB, 3.25 GB / 4GB, 3.5 GB / 4GB – and compare their performance. Details of the result of the comparison and related analysis are provided in the next section.

13.4.3.2.1 Cache Modes

Cache is the significant factor in Spark's fast performance. In Spark, or any similar data processing framework, a large amount of I/O operations cannot be avoided, and I/O operations usually cost many resources. Therefore, it is necessary to find the best mode of cache if you want the best performance. DeepLearning4J provides three modes of cache for users: default, *.cache()*, *.persist()*. Default is the moderate cache scheduling mode; it neither always persists to put a new file into the cache, nor always directly stores it into the disk. *.cache()* is the most "aggressive" way to use cache; it stores the file whenever the cache is available, and evicts the file whenever there is not enough space, *.persist()* is the most "conservative" method; it uses the cache as little as possible so that the integrity of the data can be protected as much as possible. The comparison results of three cache modes and related analysis is given in the next section.

13.4.3.2.2 Shuffle Optimisation

After monitoring and estimating the performance of each executor in the cluster, we discovered that the write function in Shuffle operation requires a long running time and a large amount of memory. In Spark cluster, an executor will create a file to store the result for each task. No matter how large the result is, the size of this file is fixed. This brings a problem; in this project, based on the statistics we observed from the profiling tool, most results of tasks cannot fill the whole file. Therefore, a large amount of memory is wasted, and the resources used for writing files into the buffer are also wasted because the system actually does not need that many files to store the results. Thus, by combining results from different tasks and putting them into the same file until the file is filled, we can significantly reduce the number of files needed to be written into the buffer, and also memory is reserved for creating files.

13.5 Performance Evaluation

13.5.1 Experiment Setup

13.5.1.1 Dataset

The dataset that we used in this chapter is called CityCam, which is collected and organised by a team from Carnegie Mellon University. This dataset contains about 60,000 frames leading to 900,000 annotated objects. The resolution of each frame in the dataset is 352 * 240. All of the images are captured by cameras placed in New York City, and the total size of the dataset is 1.4 terabytes.

13.5.1.2 Profiling Tool

The profiling tool that we used to monitor and estimate the performance of Spark cluster is SparkLens (Liu and Wu, 2017). SparkLens extends the Spark built-in monitoring tool WebUI and displays more information. SparkLens can provides information such as:

- If the cluster used more cores than needed.
- Running time and memory usage of each function.
- Minimum memory required by each worker node.

Based on the statistics we obtained by using SparkLens, we noticed that shuffle is the operation that wastes large amounts of time and memory. Specifically, the write function in shuffle is the core of the problem as we mentioned earlier.

13.5.1.3 Four Measures

The dataset is divided into two parts: the training data and the test data. The training dataset contains 48,000 frames and the test dataset contains 12,000 frames. In order to validate the test result, five-fold cross validation

		Predicted Class	
Actual Class		Class = Yes	Class = No
	Class = Yes	True Positive	False Negative
	Class = No	False Positive	True Negative

FIGURE 13.5
Four parameters.

is conducted during the test, which means, for example, as we mentioned earlier, one-fifth of the whole dataset is test data, and cross-validation means that each time we randomly generate this one-fifth of data as the test data, then evaluate it with the model trained by the remaining four-fifths of data and record the result. Then another one-fifth of the whole dataset is marked as test data and the process is repeated.

The final result is based on the results that we recorded performing this operation five times.

Figure 13.5 shows four parameters. Before we interpret the performance, it is necessary to explain the four parameters that are important for evaluating a DL model.

- True Positive: Correctly predicted positive values: both expected and actual classes are yes.
- True Negative: Correctly predicted negative values: both expected and actual classes are no.
- False Positive: Wrongly predicted positive value: expected yes but actually is no.
- False Negative: Wrongly predicted negative value: expected not but actually yes.

Accuracy is the most straightforward and intuitive performance measure: it is the ratio of correctly predicted observation to the total observations (Pokharna, 2016). In this project, it equals the ratio of the correctly classified objects to all classified objects.

$$\text{Accuracy} = (TP + TN) / (TP + FP + FN + TN)$$

Precision is the ratio of correctly predicted positive observations to the total predicted positive observations (Pokharna, 2016). In this project, it equals the ratio of correctly classified vehicles to all classified vehicles.

$$\text{Precision} = TP / (TP + FP)$$

Recall is relatively complicated; basically, it is the ratio of correctly predicted positive observations to all observations classified as "yes." In this chapter, it equals the ratio of correctly classified vehicles to the classified vehicles.

$$Recall = TP/(TP + FN)$$

The F1 score is the most complicated performance measure. It is the weighted average of Precision and Recall (Pokharna, 2016). For this project, it is more helpful than accuracy since the distribution of vehicles is uneven in the city.

$$F1 Score = 2*(Recall*Precision)/(Recall + Precision)$$

These three parts are components that we think most affect the performance of the model in this project after running a series of tests. Following are the results that we obtained in the experiments.

13.5.2 Results and Analysis

13.5.2.1 CNN Model Optimisation

Figure 13.6 shows the comparison between the default constant learning rate and the combined learning rate, the comparison between default Max Pooling layers and both Max and SoftMax Pooling layers and the comparison between the model with the default setting and the model with optimisation.

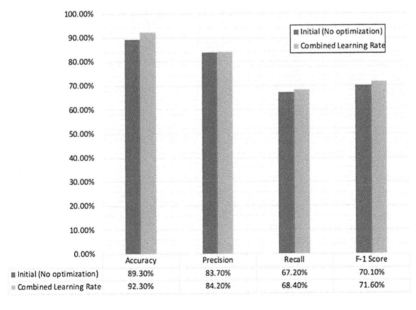

	Accuracy	Precision	Recall	F-1 Score
Initial (No optimization)	89.30%	83.70%	67.20%	70.10%
Combined Learning Rate	92.30%	84.20%	68.40%	71.60%

FIGURE 13.6
Prediction result: Combined learning rate.

TABLE 13.1

Five-Fold Cross Validation: Default vs Combined Learning Rate

	Default-Average.	Default-Std.	Combined Learning Rate-Average	Combined Learning Rate-Std.	Improved-Average.	Increased-Std.
Accuracy	0.8930	0.0074	0.9230	0.0081	3.3%	9.4%
Precision	0.8370	0.0043	0.8420	0.0044	0.12%	2.3%
Recall	0.6720	0.0132	0.6840	0.0127	1.8%	−3.8%
F-1 Score	0.7010	0.0142	0.7160	0.0123	2.1%	−13.4%

Figure 13.6 and Table 13.1 show the comparison between default (constant) learning rate and combined learning rate. Note that all the performance metrics are improved, among which accuracy is improved most significantly after implementing combined learning rates.

Figure 13.7 and Table 13.2 show the comparison between default (constant) pooling layer using only Max and the optimised pooling layer using both Max and SoftMax (as described in Section 13.4.2.3). Note that all the performance metrics are improved, among which Precision, Recall and F- Score all improved significantly with optimised pooling layer.

The performance of the model after and before optimisation on above three components is displayed in following Figure 13.8 (Table 13.3).

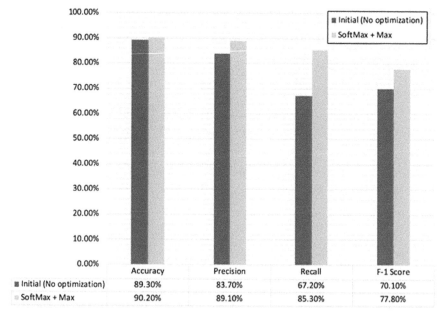

	Accuracy	Precision	Recall	F-1 Score
Initial (No optimization)	89.30%	83.70%	67.20%	70.10%
SoftMax + Max	90.20%	89.10%	85.30%	77.80%

FIGURE 13.7
Prediction result: SoftMax and Max.

TABLE 13.2

Five-Fold Cross Validation: Default vs SoftMax and Max

	Default-Average.	Default-Std.	SoftMax and Max-Average.	SoftMax and Max-Std.	Improved-Average.	Increased-Std.
Accuracy	0.8930	0.0074	**0.9020**	**0.0131**	1.1%	77%
Precision	0.8370	0.0043	**0.8910**	**0.0229**	6.4%	433%
Recall	0.6720	0.132	**0.8530**	**0.0352**	27%	167%
F-1 Score	0.7010	0.142	**0.7780**	**0.0296**	11%	108%

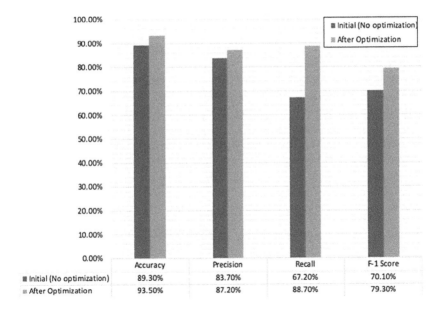

	Accuracy	Precision	Recall	F-1 Score
■ Initial (No optimization)	89.30%	83.70%	67.20%	70.10%
▨ After Optimization	93.50%	87.20%	88.70%	79.30%

FIGURE 13.8
Prediction result: Before and after optimisation.

TABLE 13.3

Five-Fold Cross Validation: Default vs Optimisation

	Default-Average.	Default-Std.	Optimization-Average.	Optimization-Std.	Improved-Average.	Increased-Std.
Accuracy	0.8930	0.0074	**0.9350**	**0.0092**	4.7%	24%
Precision	0.8370	0.0043	**0.8720**	**0.0077**	4.2%	79%
Recall	0.6720	0.0132	**0.8870**	**0.0212**	32%	61%
F-1 Score	0.7010	0.0142	**0.7930**	**0.0179**	13%	25%

Based on the test result, we can tell that our optimisation is successful. Generally, all four measures are better after the model has been optimised. Specifically, the improvement on Recall and F1 Score is great, even though the improvement on Accuracy and Precision is alright. We believe introducing SoftMax combined with Max is the most significant reason why the Recall of the model improves. The combination of different learning rates contributes greatly to the improvement of the F1 Score.

13.5.2.2 Spark Cluster Tuning

Figures 13.9 and 13.10 show the comparison between three types of memory distribution and the comparison between three types of cache modes. Each comparison is based on the four distributed modes that we mentioned in the previous section.

Figure 13.9 and Table 13.4 show the cost of time among three different memory layouts. Note that 3.25 GB / 4 GB has the best performance in all four distributed environments.

Figure 13.10 and Table 13.5 show the cost of time among three modes of cache. Note that the aggressive *.cache ()* is better than the most conservative *.persist ()*, while both improved over the moderate *default* setting (see Figure 13.11, Table 13.6 and Table 13.7).

The test result shows that our optimisation on shuffle makes a great improvement on the training time. This result is based on the configuration with best performance: 3.25 GB / 0.75 GB + *.cache()* + shuffle optimisation.

13.5.3 Summary of Performance Evaluation

In this section, we compare accuracy and cost of training time of CNN before and after using Spark. First of all, note that before using Spark, the results

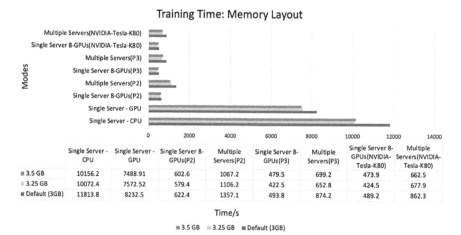

Training Time: Memory Layout

	Single Server - CPU	Single Server - GPU	Single Server 8-GPUs(P2)	Multiple Servers(P2)	Single Server 8-GPUs(P3)	Multiple Servers(P3)	Single Server 8-GPUs(NVIDIA-Tesla-K80)	Multiple Servers(NVIDIA-Tesla-K80)
3.5 GB	10156.2	7488.91	602.6	1067.2	479.5	699.2	473.9	662.5
3.25 GB	10072.4	7572.52	579.4	1106.2	422.5	652.8	424.5	677.9
Default (3GB)	11813.8	8232.5	622.4	1357.1	493.8	874.2	489.2	862.3

Time/s

3.5 GB   3.25 GB   Default (3GB)

FIGURE 13.9
Training time between three types of memory layout.

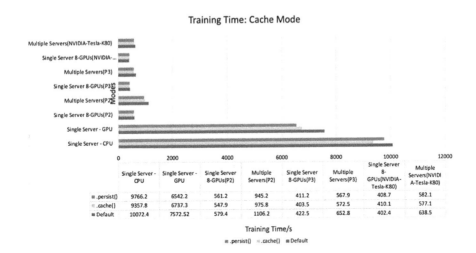

FIGURE 13.10
Training time between three types of cache modes.

TABLE 13.4

Five-Fold Cross Validation: Training Time of Three Memory Layouts (Single Server – 8 GPUs, P3 instance)

	3.5 GB-Average.	3.5 GB-Std.	3.25 GB-Average.	3.25 GB-Std.	3 GB-Average.	3 GB-Std.
Training Time	479.5s	22.3s	422.5s	20.1s	493.8s	21.8s

TABLE 13.5

Five-Fold Cross Validation: Training Time of Three Cache Modes (Single Server – 8 GPUs, P3 instance)

	.persist()-Average.	.persist()-Std.	.cache()-Average.	.cache()-Std.	default-Average.	default-Std.
Training Time	411.2s	19.2s	403.5s	21.3s	422.5s	22.6s

shown in Section 13.3.6, Figures 13.1 and 13.2, are for 5,000 frames, while using Spark the results shown in Section 13.5.2 are for 60,000 frames. Next, for accuracy results, comparing those in Figure 13.1 with those in Tables 13.1, 13.2 and 13.3, we can see that the accuracy remains high, while all the performance metrics (including Precision, Recall, and F1 Score) are significantly higher after optimisation methods are applied. Lastly, comparing cost of training times shown in Figure 13.2 and those in Tables 13.4 and 13.5, the time needed is much reduced, from the range of 837–3252 seconds to 423–494 when using CNN, and down to 404–423 seconds when applying optimisation methods.

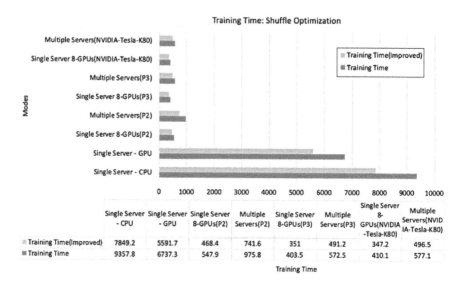

FIGURE 13.11

Training time before and after shuffle optimisation.

TABLE 13.6

Five-Fold Cross Validation: Training Time of Three Memory Distributions (Single Server – 8 GPUs, P3 instance)

	Before Optimisation – Avg.	Before Optimisation – SD.	After Optimisation – Avg.	After Optimisation – SD.
Training Time	403.5s	21.3s	351.0s	19.2s

13.6 Conclusion and Future Directions

This chapter addresses the need for speedy traffic flow prediction by discussing and presenting promising designs and encouraging results using CNN, with the Apache Spark distributed computing cluster as an accelerator. The chapter first details the background on machine learning, deep learning, and Spark cluster, as well as presenting related studies on traffic flow predictions. Some preliminary results on performance comparison between basic CNN and traditional machine learning methods are illustrated. The proposed solution, a fine design of distributed CNN running on Apache Spark cluster, is carefully described.

The performance evaluation has been carefully presented. The results of the proposed CNN model achieve both prediction accuracy and limited

TABLE 13.7

Five-Fold Cross Validation: General Improvement on Training Time

	Single Server – CPU.	Single Server – GPU.	Multi-Server – Single GPU (AWS P2).	Multi-Server – Single GPU (AWS P3).	Multi-Server – Single GPU (Google Cloud K-80).	SingleServer– Multi GPUs (AWS P2).	SingleServer– Multi GPUs (AWS P3).	Single Server – Multi GPUs (Google Cloud K-80).
Before	11813.8s	8232.5s	1357.1s	874.2s	862.3s	622.4s	493.5s	489.2s
After	7849.2s	5591.7s	741.6s	491.2s	496.5s	468.4s	351.0s	347.2s
Reduced	34%	32%	45%	44%	43%	26%	29%	29%

training cost. By deploying the proposed training model in a distributed environment, the training time has been significantly reduced while high-level prediction results have been attained.

Traffic prediction will continue to be a vital issue in many years to come. For the direction of future research, with constantly growing volumes of traffic, it is desirable to continuously work on the speeding up of traffic flow prediction while keeping computation and time costs low. While the results of the experiment presented in this chapter have shown significant improvement, they are not entirely stable based on the standard deviation obtained from the five-fold cross-validation. More intelligent designs and experiments are needed to stabilise the high prediction results and fast training time.

References

Alom, M. Z., Taha, T. M., Yakopcic, C., Westberg, S., Sidike, P., Nasrin, M. S., Van Esesn, B. C., Awwal, A. A. S. and Asari, V. K., 2018, 'The history began from AlexNet: a comprehensive survey on deep learning approaches', *arXiv preprint arXiv:1803.01164*.

Beam, A. L., 2017, 'Deep learning 101 – part1: history and background', https://beamandrew.github.io/deeplearning/2017/02/23/deep_learning_101_part1.html (accessed on May 2, 2019).

Duan, M., 2018, 'Short-time prediction of traffic flow based on PSO optimized SVM', in *2018 international conference on intelligent transportation, big data & smart city (ICITBS)*, Xiamen, pp. 41–45.

Fukushima, K., 2007, 'Neocognitron', *Scholarpedia*, vol. 2, no. 1, pp. 17–17.

Liu, Y. and Wu, H., 2017, 'Prediction of road traffic congestion based on random forest', in *2017 10th international symposium on computational intelligence and design (ISCID)*, Hangzhou, pp. 361–364.

Madhukar, 2015, 'History of Apache Spark: journey from academia to industry', http://blog.madhukaraphatak.com/history-of-spark/ (accessed on May 2, 2019).

Manzi, D. and Tompkins, D., 2016, April, 'Exploring GPU acceleration of apache spark', in *2016 IEEE international conference on cloud engineering (IC2E)*, IEEE, pp. 222–223.

Pokharna, H., 2016, 'The best explanation of convolutional neural networks on the Internet!', https://medium.com/technologymadeeasy/the-best-explanation-of-convolutional-neural-networks-on-the-internet-fbb8b1ad5df8 (accessed on May 2, 2019).

Zhang, W., Xue, B., Zhou, J., Liu, X. and Lv, H., 2018, 'A scalable and efficient multi-label CNN-based license plate recognition on Spark', in *2018 IEEE smartworld, ubiquitous intelligence & computing, advanced & trusted computing, scalable computing & communications, cloud & big data computing, internet of people and smart city innovation (SmartWorld/SCALCOM/UIC/ATC/CBDCom/IOP/SCI)*, Guangzhou, pp. 1738–1744.

Zulkifli, H., 2018, 'Understanding learning rates and how it improves performance in deep learning', *Towards Data Science*, 21, pp. 23–25.

Index

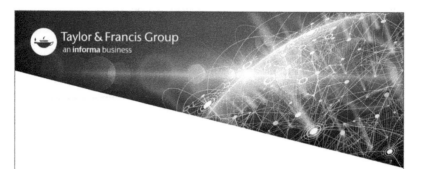